A Journey to Knowledge, Learning, and Spirit

Margaret E. Derry

HWAY

Life as Art, Science, and History

POPLAR LANE PRESS

POPLAR LANE PRESS

20243 Heart Lake Rd RR1
Caledon ON
L0N 1C0
Tel: 519.941.6894
Fax: 519.941.6583
www.poplarlane.net

Standard Address Number: 119-0148

Poplar Lane Press operates as a division of Poplar Lane Holdings Ltd.

Library and Archives Canada Cataloguing in Publication

Derry, Margaret Elsinor, 1945-
Pathway : life as art, science, and history : a journey to
knowledge, learning, and spirit / Margaret E. Derry.

Includes bibliographical references and index.

ISBN 0-9738335-0-5

1. Derry, Margaret Elsinor, 1945-. 2. Derry, Margaret Elsinor,
1945- —Knowledge and learning. 3. Autobiographical memory.
4. Painters—Canada—Biography. 5. Cattle breeders—Canada—Biography.
6. Historians—Canada—Biography. 7. College teachers—Canada—Biography.
I. Title.

ND249.D467A2 2005   759.11
C2005-903558-7

The colour plates are all the work of the author and were executed in
watercolour and in oil. They are now included in various corporate and
private collections, as well as in that of the author.

Contents

For my children, Alison and David.

Books by Margaret Derry

Ontario's Cattle Kingdom, Toronto: University of Toronto Press, 2001
Bred for Perfection, Baltimore: The Johns Hopkins University Press, 2003

Forthcoming:
Horses In Society, Toronto: University of Toronto Press, 2006
A Shadow Within a Shadow – an autobiographical novel, forthcoming

Preface

This is a true, life story that I began to write in 1983. As time passed, I would add material when it seemed appropriate. Over the last three years I devoted more energy to the project, and tried to understand the deeper meanings that my story seemed to contain. I plunged into reading about the philosophy of art, science, and history; and the history of art, science, and history. It has been a labour of love, and a mental journey into ever more wondrous discovery. It is my hope that by sharing these experiences of learning, I will make others feel their own lives in an enriched way.

I had tremendous help as I went down this pathway. My husband was the one who first proposed that my "scribbling" in fact provided the contents of a book and a good story. Both my son, David, and daughter-in-law, Shannon, read this manuscript a number of times. They are experienced writers and provided very helpful suggestions which made the book take on a more mature, and even artistic, shape. A number of academic colleagues also looked over the manuscript and offered encouragement.

Rosemary Shipton, a skillful editor, improved and fine-tuned the writing, and John Lee artistically designed the layout of the manuscript. I learned much from both of them about the production of books, my experience having been confined to academic publishing.

My heartfelt thanks to all these people. Without them and their input this would not have been the book that it is today.

PATHWAY

Mnemosyne and the Muses

O N AUGUST 6, 1982, my life changed forever. My mother died of a heart attack in the family boat, just as we started out for a picnic. Boating and picnicing were part of our lives at the cottage in north Georgian Bay, near Killarney, Ontario, and my mother had looked forward to this trip. She had wanted once more to see an island out in the bay, where strange rock formations resulted in ribbons of white silica running through the pervading pinkish-red granite. When my mother died that brilliant summer morning, I was not a child. A happily married woman in her late thirties with two children and an art career that had evolved from years of self-training – I was about to have my first gallery exhibition of paintings – I should have been cushioned from the shock that extinguishment of this life might bring. It was not to be, though. Time was severed, as surely as a knife had cut it in half, and large aspects of my life immediately became relegated to the past. I recognized that essential parts of me now belonged to that past – a fact that, in the end, would be hugely significant. The death proved to be not just a termination and a loss but also a beginning: it initiated what would become a long mental journey.

The pathway of the journey began with my wanting to hold on to my past, perhaps even bring it back, through memory. A compulsion to look in detail at specific past events filled me. Splintered visions of bygone days – flittering visual images enlarged by memory of sound, smell, and space – forced themselves, over and over again, into my consciousness. I could not think enough about my memories, could not hold more tenaciously to my past. Of course I knew this was part of the grieving process that went with the loss of my mother, but the intensity of the need to examine memory suggested to me, even shortly after her death, that something more important was going on as well. At last I found the key. Greek mythology explained what memory was trying to tell me. When I read about Mnemosyne and the muses, I realized that my memories were indeed about much more than grieving. Memory put one's feet on the pathway to knowledge. Memory was essential to all learning, even to the evolution of the human spirit. I would, as I struggled to understand the meaning of things over the years, become ever more astonished at the fundamental wisdom embedded in this ancient myth.

Mnemosyne is the goddess of memory. According to Greek mythology, she was critical to humanity's development and to the growth of knowledge in every individual as well. Humankind's ability to learn began with the birth of Mnemosyne, the daughter of Gaia (Earth) and Ouranos (Sky). Goddess of both memory and wisdom, Mnemosyne gave us language. From her union with Zeus, she bore the nine muses who lived on Mount Helicon. The muses would provide people with the tools to use language, first to gain knowledge and then to pass it on through communication.

The arts were implied in the muses' talents that related to song, poetry, and play writing: appreciation of symmetry and balance, attempts to penetrate the meaning of emotions in order to make sense of human existence, a desire to create as a result of love, and a longing for a form of beauty so closely affiliated to a sense of perfection that it must reside in the realm of the gods and be one aspect of divinity. These qualities were the territory of Melpomene (tragedy), Thalia (comedy), Terpsichore (dance), Calliope (epic poetry), Erato (poetry of love), Polyhymnia (music for the gods), and Euterpe (lyric poetry). When aesthetics joined the objective thinking found in the study of science – those efforts to comprehend the laws of nature outside the self (through Urania, who represented science in the study of astronomy) – and within the framework of historical understanding (through the gift of Clio, or history), people developed the ability to learn.

Apollo, the god of light, truth, music, and medicine, acquired knowledge with the aid of the muses, and in the end he became their leader. All of us learn in some fashion the way Apollo did, through the lessons taught by the goddess and her daughters. Here I describe the pathway of one person into the world of Mnemosyne and the muses by showing how, from an initial search for memories, I learned that aesthetic thinking is connected to an appreciation of the past's interface with the present, that science relates to art within that framework, and, ultimately, that a historical outlook binds our sense of art to our understanding of science. This, then, is ultimately a story of how art, science, and history work together in human thinking.

Happy is he whom the Muses love, sweet flows the voice from his mouth. For if someone has pain and fresh grief in his soul and his heart is withered by anguish, when the poet, the servant of the Muses, chants the fames of me of former times and the blessed gods who hold Olympus, then straightway he forgets his sad thoughts and thinks not of his grief, but the gifts of the gods quickly turn him away from these.
　　~ Hesiod, "Theogony"

I pray to Mnemosyne, the fair-robed daughter of Ouranos, and to her daughters, to grant me ready resource; for the minds of men are blind, whosoever, without the maids of Helicon, seeks the steep path of them that walked it by their wisdom.
　　~ Pindar, "Paean 7"

CHAPTER ONE

Buildings

My mother's death made me think over and over about my past. I became obsessed with it, not so much in relation to her, but rather with the past in its own right. Memory seemed to be associated with something far more important than merely a desire to think about bygone days when she was alive and with me. Perhaps my mother had given me a precious gift. She seemed to tell me, by her death, that I could not afford to lose memories. They were valuable for reasons that lay beyond their ability to retain her. And I was prepared to learn from this message. Forgetting my past, I began to see in an unclear way, would somehow make me unable to appreciate the meaning of the present. Memories, then, seemed to play as vital a role in the way life was being lived as events in the present. I no longer questioned why some memories I kept thinking about were unrelated to my mother. The need to reconstruct and preserve my past became so powerful that I started to look carefully and methodically at memories. These visions of my past would ultimately lead me into the larger implications of Mnemosyne's world. Recollections of "Poplar Point," the childhood cottage of one of my oldest friends, Sue Howard Lindell, on Lake Simcoe came first to my mind in this attempt to understand the meaning of my past for the present. I wanted, for some reason, to capture the physical layout of the place in words.

Today the cottage no longer exists. If you drive through the nearby town of Keswick and along the shore road that leads to the location, you would not be able to find where it once had been.

Everything is now subdivision. "Poplar Point" did exist, though, and in an environment that supported simple rural dwellings, a tree-lined waterfront, and small lakeside cabins. In its heyday, "Poplar Point" was a beautiful spot: a grassy property, which extended as a point into the lake, blessed with many tall Lombardy poplars, a rambling clapboard summer home, and the endless sound of moving water. The setting created vivid impressions on my childish mind which would endure into adulthood. I never hear the airy rustle of poplar leaves or the sound of a meadowlark without thinking of this spot, with two little girls playing on the sunny lawn. When Doug and I were first married, we lived on the ground floor of an old apartment building. The back window looked over a large cinder patch and a series of dilapidated garages. Beyond them stood a fine old poplar, and on still, hot summer nights when I lay restless, the endless whisper of its leaves would lull me to sleep as I imagined the murmur of the water against the breakwater shore of "Poplar Point."

The cottage at "Poplar Point" had three gabled bedrooms upstairs; a large, long living room with a central fireplace that dominated the space; a dining room looking over the breakwater; a bright kitchen; and, even more enchanting, a large sunroom, which in the off-season months contained a crazy assortment of toys. It was the summerhouse and the barn on the property, however, that I remembered the best.

Once off the veranda, a child could run quickly over the croquet lawn to the very end of the point where the summerhouse stood, guarded by aged and graceful Lombardy poplars. This small building stood on wooden stilts, forming a roof for an enchanting Victorian-type gazebo below it. The wooden floor of the gazebo, bleached and faded by many summer suns of whatever paint it might once have had, supported a quaint natural wood railing around it. The gazebo suggested olden times, with ladies and gentlemen sitting on white garden chairs sipping tea in the late afternoon sun, which would dance across the moving water that lay only a few feet away. I never saw anyone sit there and, in my time, there were no inviting chairs. The ancient wooden floor, weathered and somehow ageless, seemed to present these possibilities to my imagination.

To the left side of the railing, an uneven, wooden staircase rose to the summerhouse itself. The height was truly awe inspiring to young children, but Sue and I felt brave enough to climb the steps. At the top, standing in the warm sunshine and feeling a breeze rising from the water, we would turn to open the door to the one room that made up the summerhouse. Warm musty air, perfumed with the breath of the past, immediately embraced us. How enchanting was this small bedroom with windows on all sides! Even with them closed, we could hear the ceaseless murmur of the lake. I could imagine how it would feel with all windows open on a warm summer night, with the moonlight streaming across the floor. The furniture was simple: a small dainty bedside table painted white and a large bed of wrought iron also painted white. No one slept here now, but Sue told me that her older sister, Janet, used it as a bedroom when she was a girl.

The barn at "Poplar Point" evoked in me equally strong, but different, emotions and thoughts. The graceful driveway, which wound its way out of the heart of the point, led to the main road, and in those days the barn, with its cultivated field beside it, lay across that road. Today you will see nothing but houses where the barn and its field used to be. Sue's family had once owned a great deal of land around the barn, but in my time the barn, one field, and the point itself were all that remained in the family. The building with its neighbouring field spoke of a way of life unfamiliar to me. It was possible to imagine how people farmed the land, and how they housed their animals. I remember the smell and feel of the earth, in which vegetables could be seen growing in well-tended rows. Sections within the barn made me think of grain bins and stanchions for cows.

The old structure was a paradise for children. Long unused as a working barn, it still maintained a soul, as barns will. The barn was always cool, always dark, and drowned in a soft scent, faded by time, of wood, manure, hay, and dampness. The ground floor lay clean and quite empty. Wide doors opened at the front and a dim passageway led to similar wide doors that opened off the other side of the barn.

The floor space was partitioned; the front part appeared to have been used for cars, and earlier for horse-drawn carriages. Beyond this open area and nearer the back doors stood stalls, both open and box, for horses. Janet had once had a pony, and Mr. Howard, in his youth, had riding horses. To my mind, the idea of early morning rides on the point, amid sunlight falling on dewy grass and the lake's wavelets, seemed about as close to heaven on earth as could be imagined.

A wooden stairway rose to the hayloft. The loft had not held hay for many a year and was then used as a storage place for old trunks and furniture. It was a fascinating place for children to explore, where wrought-iron furniture and trunks of old china, clothes, and cooking utensils could be expropriated for play downstairs. In effect, we were using the material culture of the past. There was something strange and delicious about that blurring of time barriers: it was palpable to me, even if I couldn't explain why.

A strange catharsis came over me when I looked carefully at my thoughts about "Poplar Point." I felt relief. I sensed that I had resolved something by externalizing these memories. Why? Had my reconstruction somehow helped my grieving, even if it did not relate directly to my mother? Had the act of composition healed me, by allowing me to let go of the past? Or had the thinking process somehow convinced me that the past had been brought into the present? Why had my feelings surrounding the memories in effect become ones of relief?

Resolution of some sort of pain seemed to be linked to the act of depiction within the mind. Soon I realized why. My memories, my past, my history told me something that was larger than I alone. I had a glimmering that I would learn about the way people perceive their past, their history, through memory. When I thought about "Poplar Point," there was a definite pattern to the way I relived memories. I focused on buildings. I had been compelled to outline three-dimensional space and the way humanly made structures interacted with that space. Human activity had been appreciated by me within the framework of both. The simple act of depiction within the mind drew me into issues unrelated to the memories themselves. Does interest in space generally become part of the formula that makes memory, I wondered, or that translates sensations evoked by certain memories into something new? Is there something particularly important about humanly constructed space that relates directly to the power of memory?

Gaston Bachelard, in *The Poetics of Space*, provided me with some information on that problem. Interested in the effects physical space has on our psyche, he spoke of the childhood house in memories. He pointed out that the building that is home shapes our senses of the environment outside. We learn to see reality – the entire natural world – in connection with certain humanly constructed spaces. In my recapturing of "Poplar Point" through memory, I had uncovered something about this pattern of human interaction with nature. It is evident that humanly built space (even if not my home) – this cottage, but especially its summerhouse and barn – played a powerful role in the way I had related to sky, water, grass, flowers, trees, wind, and light. Nature moved effortlessly through the buildings, like summer-scented breezes through an open window. Reality of nature can meet reality of human behaviour particularly vividly in memories defined by humanly constructed space. I realized that I had lived, not thought about, Bachelard's philosophy. Experiencing philosophy, rather than contemplating it, gave an authenticity that was intoxicating to certain conceptions.

My mother's message through death seemed to release something in me: my past, my memories, my history might be able to teach me about the meaning of things for people.

~~~

When Doug and I were married in the late 1960s, we quickly found that we shared a love for the land. Many weekends we spent roaming the countryside just outside the city of Toronto, looking at land we dreamed of buying but could not afford. Over the next few years, two children were born, Alison and David. Life became so busy that, for the time being, we stopped exploring country properties. I took up my old love of drawing and painting, and in doing so I plunged myself into an exciting intellectual world. Because I had not been to art school, I read anything I could on how to paint and draw. My work improved, and I managed to sell my art by undertaking commissions to do portraits. They could be of cats, dogs, horses, children, or even houses. I produced images in both oil and watercolour, preferring these two mediums to pastel. My art did not bring in much money, but the occupation allowed me to stay at home with the babies. In the years ahead, I participated in joint exhibitions and also had a number of solo shows. Doug was busy, at the same time, building his career as a chartered accountant.

We never forgot our love of the land, and, before long, we started to look again. It might be possible, we thought, at last to fulfill our dream. Shortly before my mother died, we did manage to buy a farm, which we named "Poplar Lane Farm." While some of the richest farmland in Canada can be found in the county where this farm was located, the district of Caledon was, and is, poor in comparison. Hilly and stony, much of it has always been hard to cultivate. Settlement of land began here in the 1850s. Our farm's patent from the Crown dates back to 1856, and a small house was built here about 1875, using stones collected from the fields. A large bank barn was later constructed near the house, using field stones to make the foundations. To this day, hemp ropes and wooden pulleys rise up high in the upper level of the barn. This equipment allowed horses to power the lift of hay into the lofts. Below, stanchions for cattle had been constructed. They are still there, with rusting chains attached to the neck braces, through which the animals could reach mangers holding feed. One family farmed the property over three generations, the last of which raised Angus beef cattle. In the early 1960s the livestock was auctioned off and the place was sold to a Toronto lawyer. He and his family used it as a vacation spot. We bought the property from him.

Almost immediately, the farm evoked strong sensations of the past, of nostalgia, and of grieving in me – yet a sense of beauty, too, strangely connected with all those feelings. This complex reaction to the farm had begun to set in just before my mother died, but her death served to intensify the response. I pondered on this queer collage of emotions. Why did I associate the past with a farm we had just bought, when I had never before lived on a farm? I loved the beauty of the land – the rolling green of pastures, the arching sugar maples, the soughing of the wind in the poplars on the lane. But I could recognize this beauty fully, I realized, only through what seemed to be a sensation of nostalgia. Was beauty, then, part of the nostalgia? I looked at various definitions of the word. Nostalgia is defined as "longing for familiar or beloved circumstances that are now remote or irrecoverable" or "any longing for something far away or long ago," but it's also derived from the Greek words meaning to "return home" and to "feel pain." A wish to return to anything means also to go back in time. Also critical to the meaning of nostalgia is the sense of grieving over loss, but it seems clear that the pain carries a sense of beauty through love. We do not feel nostalgia for something we do not love or hold as beautiful through that love. I wondered whether I could define nostalgia as that form of memory that perceives aspects of the past with an appreciation of beauty and love in the present.

I felt intense nostalgia when walking into the barn's dim musty coolness. Why was it so disturbing? How could our barn provoke nostalgia without some real base for memory? Or was there memory? And memory of what? Of course I recalled the "Poplar Point" barn and heard again the music of poplars near the lake. But surely that barn at Lake Simcoe could not be responsible for such powerful feelings. I wanted to understand more about the power of nostalgia, especially its characteristics of love and beauty, in relation to physical space. I wondered if these qualities had anything to do with its apparent ability to blur past into present. A humanly constructed space, this time our old barn, served me in my exploration of these ideas.

I thought about any barn memories I had. One concerned a barn on a farm near my childhood friend Dale Robinette's place at Southampton, Ontario. Several times Dale and I had played at this farm near the Lake Huron cottage. I remembered the fieldstone farmhouse and the wide pastures in the late summer afternoon sun, with cattle grazing peacefully in them, and a broad sleepy barn at the edge of the fields. The farm provided a physical setting for the barn, which was usually the focus of our attention. How we loved playing in its hayloft, where sweet and soft hay was made golden by thin fingers of sunlight filtering through cracks between the boards. I still remember the warm scent of sun-bleached wood, of hay, and of dust in that particular barn. We ran and jumped in that glorious loft until we were hot and tired, and then we went out to the cooler pasture to see the cows. We wandered over the fields that seemed so large and full of tall grass. I feared the cows might charge us, but they never did.

Is the barn in Southampton the one I am thinking of when I enter our barn and look up into the decaying loft shrouded in ancient hay? No, I thought, there is something else. Best to look even deeper into my past. One barn came to mind quite easily. When I was about four years old, I visited the farm of Mother's aunt. Even though the farm fields are long gone now, from the expansion of the highway, I still remember the old place well. After dinner, my father took me outside. We walked from the house east to the barn. My father held my hand and led me, with the evening summer sun warming our backs. The wide doors stood open, and I could see the barn's dim interior. We went in. Faded memories of animals in the gloom and of the scent of fresh hay return to me. But that memory carries little nostalgia.

Something in my mind began to happen. From the depths came other memories that I knew had vital meaning. And then I realized that the nostalgia evoked by our barn was, in fact, rooted in real memory. I came at last to the place I recognized to be the memory I had been trying to find or relocate. I knew what I had been seeking in my past with respect to this farm and this barn. I understood, too, why I had always wanted a farm. I wanted to recapture life rhythms found on a property on the main street of Brampton.

Brampton was where my mother's old home had been. The house still stands today, but it is now so surrounded with other buildings – houses, a school, and a church – that the setting I remember has gone forever. I can see in my mind's eye the old property as it was years ago – the wide sunlit lawns, a large Victorian house set among tall trees, the rose garden, the pond, the corn fields, the orchard, and the red and green barns. I have memories of Brampton from before my grandmother died, when I was five years old (my grandfather had died before I was born), and others that date from after the time she died. Several years before her death she had sold the family home and built a new, smaller one for herself beside the old house. But she kept the property – the fields, barns, orchards, and rose garden. Here we would come from Toronto for dinner. I remember one spring evening, when we stepped out of the car to hear the fluttering song of robins, how excited my mother was as she showed her mother a new diamond ring. I wondered why everyone looked so interested. The stone seemed too small to be of any significance. I knew of much larger and more beautiful rocks of translucent silica that could be found at our family cottage in Killarney!

Early recollections of the Brampton property fused together to form one overarching memory, which could be characterized by the sensation of prevailing love. From the enfolding warmth of that love, I saw and felt the larger environment around me. There was the sound of a train at dusk beyond the corn fields, the beauty of the roses in the garden and over the white wooden trellis. The creek that ran through the property flowed into the centre of town, and I remember it across from the park. When I spent a weekend alone with my grandmother shortly before she died, we went to hear a band in that park and to a fair where she bought me a doll with pink feathers. I do not recall her face, but I remember the person because she personified love.

The Brampton memories after her death were those that revolved around a barn, and they ultimately tied barns to all the sensations around Brampton. After my grandmother died, Mother sold the little house but kept the property for a short period of time. We used to go out on Sundays from Toronto, and my parents fixed up the green barn with cushions and garden furniture. We would take these things outdoors, never using the barn really as anything more than a storage place. These are my clearest memories of Brampton – and how I loved it! The smell of the long grass and corn, the heat, the flies, and the stony walk down by the creek. I found a flattish stone there, cracked in half, and I felt it looked like a loaf of bread. We often put up a hammock, overlooking the fields, and there we could drowse away hot afternoons, bothered only by flies. I used to wish that time would stand still, that those moments would last forever. But, with a piercing sadness, I knew that soon we would get into the car and return to the city, leaving the barn to dream in its loneliness and to hold its impenetrable secrets about the past.

The barn was full of odd things to play with. An ancient wheelbarrow, sawhorse, strawberry box, bit, and rope made a horse and carriage for my dolls and me. The loft might have been empty of hay, but it was still fun to walk on and dream. The height, the smell of ancient wood, and odd pieces of harness made it a treasure box of the past – one that I even then unconsciously recognized as being part of my own. No wonder the barn in Brampton presented agelessness to me, a heritage that later came to mean beauty through nostalgia for that barn.

By means of these attempts to understand memories, I had more understanding of how meaningful buildings could be – and not just buildings we are familiar with, such as the childhood home or cottage. I could not go back to the Brampton barn, but it could heal me through other barns. Buildings new to us can provoke nostalgia, with its abiding sensation of love and beauty, from hidden memories. There is something quite strange about experiencing what could only be described as aesthetic – an appreciation for beauty – in an environment where it is least expected. Perhaps nostalgia from hidden memories explains why. My memories and history had now shown me that, while I might understand my past in the present through buildings, that past shaped my ideas about beauty and art. I wondered if this pattern was not only mine: the linkage of aesthetics to memory might be universal. Memory and the past could be personal or collective, and probably both went into the moulding of human thought. Mnemosyne and the muses seemed to work on a double level. They might allow people to use their own past and history to understand art, but the goddess and her daughters worked together to wire the human mind. Understanding their gifts meant perceiving how memory could interpret the past and synthesize beauty from historical happenings.

~~~

About 1906 my mother's family built a cottage near Killarney. This small village in north Georgian Bay lies on a protected but deep channel of water. It is beautiful here. On one side of the channel, the pink granite rock of an island clad in green pines rises high above sparkling water. On the other side lies the mainland, where the shore is lined with huge willows and grassy spaces, amid the docks and buildings of the village. As early as 1837 Anna Jameson, a well-known English writer and the wife of Upper Canada's attorney general, described the channel's setting with such clarity that it is possible to identify the exact locations where she saw campfires and native encampments. This place has an ancient history. Formally named Killarney in 1854, the area had served as a settlement for the Ojibwa for centuries. As the nineteenth century advanced, the village became a centre of commercial fishing. Lumbering developed near by – at Collin's Inlet, for example – and shipping increased. Steamers regularly went through the pretty channel on their way north to Manitoulin Island and the Lakehead on Lake Superior, or south to Ontario's heartland cities. Boats docked at Killarney to unload supplies and to take on tons of whitefish. It was a busy place by the end of the century, but it was not cottage country.

"Grandpa" and "Maimie" Bell, my mother's grandparents, who lived on the Ontario side of Lake Erie, took steamer trips along the Georgian Bay shore for several years around 1900. They fell in love with Killarney and the beauty of its surrounding country. They decided to build a summer home on the George Island side of the channel. With no road access to Killarney (there was no road until the early 1960s), they had to bring everything they needed to build and furnish their house by steamer to the town, and thence across the water by local fishing vessel. The small cottage they designed had one room and an outhouse. There was a tiny kitchen, but they had most dinners at the hotel in the village. In 1920 they built a small fireplace to replace a wood stove for heat. Electricity came after 1950 and running water in the late 1960s. It never had gas.

When my mother was a child, she spent many summers with her grandparents in Killarney, and she would eventually inherit the property – thus it would be my childhood cottage. Killarney was the love of her life. The fact that she died there, that she loved it, and that it was so full of my past with her made it a difficult place for me to think about. But I often felt compelled to do so, especially when I was not actually there. Killarney's past could flood my mind. As memories engulfed me, I focused again on the physical aspects of buildings.

It was as though the years fell away and I could walk again as a child along the moss-covered cement path beside the cottage wall. The wall and the walk, as they were when I was a child and as they still are now, always seemed to hold the souls of people, newly gone or long gone, who lived and loved there. Somehow I could almost touch and smell the wood. Vision in the mind's eye of this mossy path triggered powerful nostalgia: I seemed actually to be living in the past again. With nostalgia came that twin sensation of beauty and love.

Contemplation of these aspects of a specific building – the cottage wall and the cement walk that ran beside it – opened the floodgates to splintered sensations of life as it was for me years ago. The sense of bleached heat on rocky picnic islands. Heat and space that drown time. Endless blue of water and sky fading into each other at the horizon, seen beyond the smooth pink rock of the shoreline. Mother, walking along in front of me, carrying a bird feather. The sense that the only reality was that moment, a delirium with no beginning or end, the knowledge that there simply was no room in your heart for pain or distress. Other memories crowded into my mind, like flitting moths. The doorway open, allowing the entrance of all the sensations of late afternoons into the old home cottage: gull calls, boat sounds, and flooding sunshine. The smell of Mother's baked potatoes and roast beef in the oven, and the sun's warmth stealing through the south window, dappled with shade from the old trees. I thought of the walk down the path, peppered with acorns, to the outhouse, the sound of the steamship *Norgoma*'s blast as she entered the channel from Little Current in the early evening.

These memories brought back to me the words of Henry Williamson in his *Flax of Dream* series:

> From the memories of those taintless days I draw my strength; by the past, man's mind is made strong with beauty. My brain is a forge of fire where from the precious metal of those taintless days, I hammer out my images. Sometimes under the stars I feel my love is with me, my love lost forever, and misery quenches my mind-fire, and the images are corroded, and all is illusion; and I am alone, on an alien planet … Could I but tell you of the glamour of those fled springtimes, when the meadow grasses waved their plumes in the wind, among the ox-eye daisies and the sorrel. Or the nightingale singing in the copse at night! … I pray for power to bring back the awareness into the human mind; I feel in my mind all the flowers and the songs of boyhood are stored, and I must pour them out, giving them shape and in sentences which will ring in the hearts of all who read, and soften them, and bring back to them the simplicity and clarity of the child-heart.

Williamson described the pain, the beauty, and the love that nostalgia provokes, but he also suggested that nostalgia can be a teacher, and that it seems to work through the sense of love and beauty it arouses in us. I had begun to suspect the same thing. Childhood years create the fabric out of which all later thought will devolve. We continuously remould the present into the terms we know and often love. Without the past, we could give the present no heart. The reasons to feel and the tools with which to feel are created at the beginning of life. And we use and remodel information of delight later in life with these tools. Nostalgia, then, helps form the way we think.

Nostalgia is obviously important if we want to understand the meaning of beauty. And I wanted this knowledge because of my deepening concern with art. On the next trip to Killarney, I was anxious to feel the beauty of this nostalgia, and not the pain of loss. The physical presence of the walk and the cottage wall, I thought, would make me sense the beauty of the past and the strength of its meaningfulness even more powerfully than the mere contemplation of them. Something strange happened, though, when I went to the old home cottage. I walked along the mossy cement at the cottage wall, but I did not feel what I sensed in my visions while at home. I tried. I would walk the path but realize as I stepped onto the veranda that I had not felt the compelling beauty that nostalgic memory could provoke. That puzzled me.

Then something really significant hit me. Nostalgia, and probably the most powerful form of it in relation to the sensation of love and beauty, seems to be triggered by the mind alone, not the physical environment. I could, I realized, experience two types of nostalgia with respect to Killarney, one stimulated by being at the place, and one that lived only in the mind. As I stepped onto the cement walk, I had failed to see that two distinct nostalgia forms existed: one form could intermingle with the other, and each could also stand on its own. I could go to Killarney, see the cement path, and imagine the people there as they once had been. But there was no other physical place from my past that I could revisit. Brampton, the barns there, and even "Poplar Point," I would never see again as they were in the past. It was only the Killarney cottage I could see today much as it was yesterday. Little, from a physical point of view, had changed. As a result, it had been difficult for me to understand that it was the nostalgia in the mind's eye which provided the most aesthetic sense of the past. Nostalgia, a form of memory that arouses sensations of beauty and love in relation to the evolution of time, inspired me when it lay in the mind only. I called this form of nostalgia "aesthetic nostalgia" because I realized that it made me want to create art – images that portrayed the past in terms of symmetry and balance, love and beauty.

I don't think this idea is new – that the nostalgia used in creativity lives only in the mind. Writers like Marcel Proust, who dwelt on the meaning of the past in the present, did not seem to want to return physically to the location where the events actually took place. Creative or aesthetic nostalgia is a process of the mind which evolves from a crucible of memory. Mnemosyne, not the work of scholars, taught me this reality. The very fact that I learned to understand the process of nostalgia in this way struck me as important. Discovering knowledge through the muses and memory allows for greater appreciation of what others have to say on the same subject.

Luckily I could test my theory by revisiting the old cottage and then seeing what type of nostalgia I felt. I went back. As I walked again over the mossy cement walk onto the veranda of the old cottage, I could feel, in the stillness, myself as a child swinging in the hammock in the old grove or playing on the rocks above the dock. I could see my father reading on the veranda in the mornings and late afternoons. And Mother sitting inside by the east window, wearing the old brown sweater and knitting her bedspread. And I remembered my father's parents crossing the walk from the sleeping cabin, Dad's mother in a flower-print dress and his father in a white suit. When I looked at the old ice box, I could sense the presence of other old people – the Bells who built the cottage, walking the cement walk, Mother's Maimie Bell in stiff, full-length dresses and her Grandpa Bell in a three-piece suit with his watch chain. But this was nostalgia evoked by a certain location. It was different from the nostalgia brought forth by the mental vision of the cement walk. I had experienced nostalgia triggered by the physical environment.

Death and grieving made me revisit memories. I came to see that this invasion of my past into contemporary existence via memories was related to something other than grieving. It was related to humanness itself. Mental descriptions of my past quickly brought me to Mnemosyne and her daughters. I learned about aesthetic nostalgia through experiences surrounding buildings. My memories informed me about the world of aesthetics, and I came to see memory generally as part of art. I began to think about memory and, therefore, the issue of time in relation to art. I was increasingly interested in those daughters of Mnemosyne. I wanted to know more. The idea that I could learn about the human spirit by looking at my own memories had taken root in my mind. My journey into understanding the importance of the past to human thinking, though, had only just begun.

Zeus created beings who would use words and music to beautify every-
thing he was responsible for: and these beings were the nine Muses.
They brought humanity the purifying power of art, which could be used
by all, not just artists. Art provided the type of wisdom, for example,
that good judges needed when they appreciated precedent. Memory lay
at the bottom of the beauty and wisdom the Muses taught, because
Memory was the mother of the Muses.
   ~ Ancient Greek myth

CHAPTER TWO

Collage

The ancient perfume of our old barn still permeated my soul, but I now understood why. The old cottage at Killarney had encoded my mind in a critical way, and I knew why this had happened too. I had learned as well that memory was related to mental sensations that were aesthetic in nature – a feel for beauty and symmetry, for example, which aroused a pleasurable appreciation that could only be described as balance. It was nostalgia that was at the seat of such powerful memory. Could I put these aesthetic internal feelings into visual images? Or, in fact, did I want to create visual images simply because they made sense of my nostalgic memories? I knew that art arose from the distillation of nostalgic memory. But nostalgic memory derives its power to produce aesthetics, I saw, from what might be described as different types of collage.

One form of memory collage in art works through a fusion of the painter's and the viewer's memories. I knew that memory could be triggered by apparently unfamiliar things – buildings and landscapes I had never seen before. If I poured my memory into art, I thought, I might be able to make viewers find personal memory in the images meaningful to me. If I succeeded, they would feel beauty through nostalgic memory encoded in their minds. Viewers did not have to understand *my* nostalgic memory to "read" my art. The old barn at our farm had taught me that I did not even have to understand where my own nostalgia arose from in order to feel its powerful presence. When away from the farm, long before I had learned to associate the building with Brampton memories, I used to dream of its soft musty interior – so cool and dim on a bright, hot summer afternoon. I knew it was meaningful, even if I did not know why. I knew I had to get back to see it and to smell it again. A sweet bitter pain, one of the sharpest and strangest of forms of aesthetic feelings, became associated with that barn in my mind. A strange blend of pleasure and hurt can reside in the beauty nostalgia provokes, and no more so than when someone looks at something apparently never seen before. When the source of nostalgia is shrouded in mystery, our response can be surprisingly intense.

Another form of memory collage felt in paintings results from the artist's sense that time passage and memory represent an overlapping phenomenon. Memory brings the past back to the mind. But nostalgic memory might have the power in art to do more – to make us feel our past as part of the present, or even to make the past collaged to the present and to the future as well. These thoughts were tantalizing, and I decided to study them through acts of painting.

I produced pictures that depicted the present-day atmosphere of our farm, where I found endless subject matter in skies and fields, wind and trees, cattle and people, in the hope that images based on them might, through a collage of vision, even bring back buried memories of a heritage: our historical agricultural past. I used my children at Killarney as subject matter for art, too, and hoped to capture something of what summer can evoke in people. Don't we often think of summer not as a present but as a past, a childhood past, coloured with nostalgia? Isn't summer, at least sometimes, and especially in cold northern countries, associated with memory that is pure beauty: love, warmth, fun, and peace? As I thought about these things and tried to understand visual reality around me in a multifaceted way, I thrilled to the recognition that I was learning to draw and paint better. I found myself discovering art as a complicated, layered thing: a problem of technical skill, visual skill, thinking skill, and emotional skill. I felt like an explorer and sought out anything that would make me a better painter and, concurrently, a better thinker about the nature of aesthetics itself. My learning evolved by living or experiencing art as an all-encompassing development, in the same way I had "experienced," rather that "learned," Bachelard's philosophy. The effects on my mind were just as intoxicating.

The more I painted, the more I became convinced that meaningfulness in a visual image often results from memory as collage – the artist's and the viewer's working together, as well as the past felt in the present. But there seemed to be another type of memory collage palpable in the art as well. Memories, not from one time but from across time, appeared to have an aesthetic effect in the paintings. I then realized that memories could infiltrate each other, and, when they did, they had a most powerful affect. I decided to study this new idea of memory collage more closely. I used memories from my past to create an image composed from separate visions that emerged over my life at different times and locations, and then to assess my reactions. In order to express time as a collage this way, I changed mediums. I used technology, not painting. When I reproduced images on a computer, and with a scanner, I saw again the powerful impressions that memory could have on the mind. With a scanner and a computer, I found I could create unified pictures that provided new images that were based on diverse splinters of the past. The images were as "old" as they were "new," because they resulted from what might be described as transformed memory. I could guide memory as well. In a sense I was harnessing memory fragments to encourage a deeper sense of the lack of concealment. The wholeness of the larger transformed memory evoked a profound sense of perfection in me because it seemed so complete. There was something almost visceral about its impact on my thinking and feeling.

The wholeness was made possible because the created image resulted from a composition made from diverse pieces of transformed memories. The splinters were unrelated to each other for anyone but me. Photographs of a place or a person could be put with pictures of something that, for some reason, merely felt like a related memory. The scanner covered these various images, and when the printer yielded a resulting picture, the physical page united the images in a way that gluing them together would not. Let me explain by describing one image of a transformed memory emanating out of the Brampton past.

I took a small picture of my mother standing in the garden in summer. She was young when the picture was taken, and it was long before I was born. Beside her I placed a small and well-designed photo of my grandmother sitting on the veranda of the old home. On a table beside her was a bouquet of flowers. She was reading. Beyond her you could see the lawn and trees. I added a photo of the old Brampton house itself, showing the place drowsing in summer sun. Surrounding the building, as if they loved it, were old and tall pine trees. I then added an image in colour that I had found somewhere of the interior of a stone farmhouse. The viewer is looking through a window at a rolling meadow and an old barn. The deep walls of stone are very evident. An old pincushion sits on the sill, and a fly swatter hangs nearby. A jar of preserves stands on a shelf. The window is open and a calendar tells the viewer it is July. My last addition was the coloured reproduction of a pastel image done by the artist Lucy Kemp-Welsh of a splendid workhorse pulling what must be a plough (but this is unseen) under what looks like a breezy, Ontario, April sky, though she was painting a British landscape, so it probably represents March. The field is quite bare, but it presents space and air. You can almost hear a meadowlark and smell the newly awakened earth. I scanned the images, and then unified them by printing them out on a common page.

When I contemplate the collected image, it fills me with a sense of satisfaction so powerful it can even change my heart rate. Here Brampton, my mother, and my grandmother became meshed with our farm. The stone farmhouse at "Poplar Lane Farm" is like the one in the print, but its jam jar, fly swatter, and barn were from Brampton too. The workhorse is a seamless image of both places, because I saw workhorses at Brampton and I know the splendid beauty of the April fields here at the farm. Because the image of the stone house stood on the wall of our cabin at Killarney, I feel a relationship to that place.

The combined image has transformed memories, by fusing them together, into a daydream, and it could become the forge for countless poems or for countless paintings.

This collaged image showed me how memories could work together across time. And then I realized that my acts of creating images, by painting and through technology, had catapulted me further into wanting to know more about how memory encodes our minds. I saw that older memories could play a role in the creation of newer ones, for example, and the present could remake our perception of the past through either new or old memory. A sense of the present can almost slip away, when memory pervades the mind and heart. I am not alone in experiencing this phenomenon. When you watch people consumed with sensations of the past, they almost seem to have left the present, and they also appear to be experiencing a profound peace, one with a mental and physical basis. It's almost an aesthetic sensation, really, with that sense of time suspension. I couldn't define what was aesthetic about it, but it appeared to be related to an appreciation of wholeness that the deep contemplation of the past within the present seemed to suggest. I had seen these effects of memory on other people. Mother often talked of life in her old home, and once, as I saw her feel her past within the present, I knew that peace and beauty dwelt in her mind. She was walking under the sugar maples, flaming in autumn gold, at our farm. She seemed so content, and I knew she was thinking of Brampton, and perhaps also of her long dead parents. Memory, for her under these conditions, became suffused with aesthetic thinking. I had seen her feel like that only in Killarney, where she had spent so many happy childhood summers with her grandparents.

As I contemplated my mother's response to our farm, I also came to see that compelling memories were often about a past before our lived past. I began to wonder if stories told to me, but not experienced personally by me, could in fact have as strong an effect on me as recollection of events lived through. By word of mouth, other people's memories had been passed on to me, and these stories often became almost as vivid in my mind as my memories. Memories could link time before my existence to me. I reflected on what had been said to me about the family in times before I was born. My grandmother had grown up in a large country home on Lake Erie. I heard so much about this place from my mother that I could visualize my grandmother riding sidesaddle on her horse, Bonnie, and imagine the sense of country stillness. I had seen the old home, large and rambling, still guarded by its wrought-iron fence, only once.

My sense of life on that property had been made more vivid by a short diary that my grandmother wrote for me while she was building her new dwelling at Brampton. On December 23, 1946, she wrote: "My mind goes back continually to the times when I was a little girl, playing in the snow with my sleigh, and Christmas day was so exciting, coming down in the morning, before daylight, and there were such wonderful presents. And the night before was the Xmas tree in the church on the bank, at Oxley. There were presents there too, and a piece to say and a part in a dialogue. Now even the church is gone; there will hardly be a sign of where it was, but someday you may stand there looking at the lake and see in imagination two little girls in white dresses, gold lockets and chains, Leghorn hats with long silk streamers, sitting in the summer shade of a big tree that may still be there by the side of the bank, waiting for Sunday school to begin." Another entry, dated March 21, 1949, read: "The first day of spring, and sunny and mild. The first spring in my extension of time as it were. I am still thrilled with watching the gulls flying over the creek, floating silver in the sun, and the willow trees that have become very yellow. Seventy springs have come and gone for me; those I remember best were the warming days in April and May in Oxley, with the delicious sensation of cool cotton underwear …. in that spring long ago, I still feel the cool dampness of the earth, as I walked bare-foot in a furrow behind a man plowing, having been allowed to remove my shoes and stockings on my return from school."

And then I saw collective memories, such as national memory, differently. Nationalism is based on memory. And nationalism is also a form of oral memory.

~~~

34

I always loved animals. As a child I had played with toy barns and animal figures. I used to provide pink lemonade for my charges by fitting a pink candle into a toy trough. Now that we owned a farm, I thought more and more about the possibility of having farm animals – perhaps not too sensible an idea, given that we lived and worked in Toronto. But still, it is hard to put one's dreams to rest completely. As Alison and David came to know the country better, they too wanted farm animals. The three of us became particularly interested in cows. Why couldn't we have cows, or at least one cow? we said to Doug. We had a farm, so why not cows too? Doug did not share our enthusiasm. With good sense, he saw problems: expense, commitment of time, as well as logistical issues around feeding, watering, and providing shelter for the animals (we all agreed that our old barn could not be used this way). Alison, David, and I persisted in our request for cows, and we claimed we could overcome all those difficulties.

One day, as we sat outside under the trees eating our lunch, the three of us brought up the subject in earnest. Doug laughed this time, but also relented. Okay, he said, but you three must do all the work of looking after them and decide on the correct animals to buy. He would help by looking into proper fencing and shelter. The children and I were ecstatic. At last, cows! Over the next months we visited many farms and met lots of bovine friends. Finally we settled on four purebred heifers that were to arrive at the farm in late June, to time with our arrival at the farm for the school summer holidays. And then, before we knew it, the great day had come. Our four girls shuffled off the truck that delivered them and roamed out into their new field, lush with tall spring grasses. The animals seemed to feel at home, even though we worried about issues such as long grass making their eyes run. We fussed as any new parents would. It was the beginning of something that would change all of us – that is, the humans. At times, we must have appeared to the cows to be slow learners, but over the years the animals managed to teach us much.

In fact, one of the greatest happiness that "Poplar Lane Farm" brought to the family was a new ability to associate with and raise animals. We have now bred generations of purebred cattle here, and every year brings with it the promise that none of us can resist: the process of life's renewal found in newborn calves. But, as I watched the youngsters frolic on spring evenings, I slowly came to realize that the pleasure they aroused in me went deeper than that. The calves and their mothers presented in living, visual form both memory and time as collaged systems that are intertwined with each other. The desire to breed animals seems to stem, at least partially, from the pleasure people derive from sensing time as a collage. Cattle reveal in living form the way passing generations can show one aspect of continuity in the passage of time – sameness. Calves born today often bear resemblance to their grandmothers, great grandmothers, and even earlier relatives. It is the most purifying thing to watch a wet newborn calf struggle to its feet, lifting a small face that looks like a long dead ancestor. The feelings provoke aesthetic sensations: time has been presented as unity by the calf, a living unity in front of me. Time, in fact, has been proven to be a unity through the truth of my vision.

Our cattle are purebred: they carry registered pedigrees. The whole culture of pedigree keeping fascinates me because it is related to so many issues: breeding standards, evaluation of stock, and issues of exclusivity. But record keeping for animals is about more than that. It is about us and our perceptions of reality. Pedigrees help us see the union of the past to the present, because the living animals can be linked to the past by well-kept records. Fundamentally, pedigree keeping evolved from the belief that the past is important to the present. Pedigrees make the past become part of the present, and they also suggest what the future might bring – all of which, of course, implies that time can be unified. Pedigrees show that time is linkage.

We have bred Murray Greys (it is these animals that appear in the paintings) and Shorthorns. The latter, in particular, lend themselves to historical study. Shorthorns are the oldest purebred breed. Their breeding records have been documented for close to two hundred years, and animals living today can be traced back to ancient ancestors – surely one way to bring the past to the present. Of course, preserving aspects of the past within the present does not mean the same thing as bringing the past into the present. One young animal on our farm is a Shorthorn heifer named Ruby, and her pedigree fascinates me. Ruby, so beautiful and alive, a splendid heifer of a rich red colour, has a pedigree that indicates she carries blood that was created by a master breeder. The heifer's full registered name, Scotsdale Ruby Broadhooks, links her to Broadhooks, a calf born to a cow named Eliza, purchased in 1844 by Amos Cruickshank of Aberdeen, Scotland. Linkage makes for continuity.

Living calves can show how continuity reflects sameness. But when pedigrees are related to the animals, calves can also show how continuity means not sameness, but change. Ruby represents documented continuity through her pedigree, yet she does not resemble her ancestor Broadhooks. It is the continuity in time passage, an issue above sameness, that in this case is valued by me in both the pedigree and the calf. Connection of past to present may imply continuity, but that is not synonymous with sameness or lack of change. Perhaps we are inclined to see one as the other – continuity as no more than sameness. Perhaps pedigrees interest us because they show us that links to the past can exist without sameness.

While purebred breeding might be described as a hybrid activity that arises out of art and science, its intimate relationship to perceptions about the meaning of time plays a role in its essential popularity. Aesthetics in animal breeding can be found in the desire to create beauty, but also in the desire to do so in relation to the flow of time. Breeders seek both to perpetuate beauty and to enhance it over time. The drive for art in purebred animal breeding is the drive to appreciate beauty in relation to time. Animal breeding is, of course, about science, too, because it deals with the process of biology and does so within the framework of time passage. Genetics itself is surely about the effects of time on all living things. And, in fact, it can be difficult to separate concerns with domestic animal breeding (artificial selection) from those of evolutionary biology (natural selection), partially because both assess life and change over time.

Ruby's pedigree made me think about certain aspects of human culture: public archives, which hold papers and manuscripts that relate to the human past. A large part of the records held in these establishments relates to genealogy, and most researchers frequenting archives are looking for their own past, not for evidence needed to support historical research. They look for their own pedigree. Why do they do it? Scholars have often denounced the whole process of pedigree keeping, on the basis that elitist thinking drives it. There is no question that elitist thinking is one facet of it. Purebred breeding is partially derived from notions about the preservation of purity via pedigrees. Human pedigree-keeping can take us into the shadowy world of eugenics, the desire to purify the human race through the elimination of "defective" people. Is elitism, though, the main or only aspect behind pedigree-keeping? Are genealogists simply elitist? No, I don't think so. They care about their pedigree because they are driven to unite their past to their present and, possibly through this avenue, their future. They need to understand time and its linkages within the framework of their individual lives.

I had felt time aesthetically when I contemplated our barn and aspects of the Killarney cottage. I had explored the artistic effects of collaged memory in visual images. I had become aware of the influence these ideas could have on certain activities, more particularly on how we interact with the process of animal breeding. Suddenly it seemed that practically everything that impinges on our lives could be perceived through a form of aesthetic collaged memory. Horseshoes, for example, could be just as meaningful as buildings. Horseshoes, especially old ones, mean good luck. We see them or models of them in shops. We know what they are and we value them as luck bearers, even if we don't have horses. In this circumstance, the shoes usually act as symbols, or cultural texts, but provoke little emotional sentiment. On this farm we find horseshoes buried in the earth, many near the old barn. The shoes are tangible objects that bring the past to the present. When I find a shoe – heavy with rust, a large one for a workhorse, and half buried in the soft earth – I am overwhelmed with the sense that the object carried a hidden story and, with that story, vital messages for the present from this farm's past. This sense triggers an emotional response, and I feel the presence of beauty.

Sound can be as evocative as vision in collaged memory: a robin whistling in the evening spring light on green lawns, made emerald by spring growth and the lingering daylight's last slanting rays from the sun. I feel suspended when listening to the sound – it is yesterday's spring, it is today, and it has a special sweetness because of the conviction that it will be heard in springs to come. The sound collages time in the mind and fuses that collaged effect to beauty through nostalgia.

I began to recognize that passages in literature appeal to us when we can relate them to our appreciation of time as a collage, even if they may not be designed to explore the effects of time's movement on our psyche. Ellen Glasgow's words, at the end of the novel written in 1904, *The Deliverance*, serve as but one example:

> With the first sunlight he awoke, and, noiselessly slipping into his clothes, went out for a daylight view of the country which had dwelt for so long a happy vision in his thoughts. The dew was thick on the grass, and crossing to the bench, he sat down in the pale sunlight beside the damask rose bush, on which a single flower blossomed out of season. Beyond the cedars in the graveyard the sunrise flamed golden upon a violet background, and across the field of life-everlasting there ran a sparkling path of fire. The air was strong with autumn scents, and as he drank it in with deep draughts it seemed to him that he began to breathe anew the spirit of life. With a single bound of the heart the sense of freedom came to him, and with it the happiness he had missed the evening before pulsed through his veins. Much

yet remained to him – the earth with its untold miracles, the sky with its infinity of space, his own soul – and Maria! With her name he sprang to his feet in the ardour of his impatience, and it was then that, looking up, he saw her coming to him across the sunbeams.

The sense of time's significance and of the resolution that results from the unity of time breathes in a hidden and subtle way through this passage. The theme of fluidity, continuity, and linkage of time pervades the writing so subtly that I wonder if the author recognized how strongly they influenced her thinking or had been encoded in her humanity. "Field of life-everlasting" suggests a space that holds all time – past, present, and future. Across a space of all time, and perhaps binding that space and time together, "ran a sparkling path of fire." Personal past melts into eternity in a hidden way – for the bodies of his ancestors lay nearby in the graveyard protected by the cedars. The air "was strong with autumn scents," fragrance of the immediate past, and he "drank it," in order to pull himself into the present and, in the process, "renew the spirit of life." The vividness of the present (a single rose out of season, the freshness of the morning seen in dewdrops) triggers energy and anticipatory happiness, because in the coming of the future "much remained to him … the sky with its infinity of space." All meaning and time were personified in Maria, and "he saw her coming to him."

Time is linkage, and linkage is continuity. But this continuity presents a strange ambiguity, because sameness and change are both aspects of it. Time passage entails a continuity that can express both sameness and change. Static states may also exist over time, and changing states may have static components as well. The movement of time, then, means fluidity with change as much as with sameness.

Now that I had grasped how significant these ideas concerning time, memory, and nostalgia were to the workings of the human mind, I began to take greater notice of what other people thought about these things. I was more interested, for example, in perceptions about time in the play *Berkeley Square*, written in 1916 by John Balderston, than I might otherwise have been. The author believed that a person can see past, present, and future as three separate entities or can see them together as part of a continuous unit. Balderston explained how two people could view the same passage of time differently. One man watched a river landscape from a boat. He could see what was beside him on the river bank, and he knew what he had passed by, but he could not know what the future would bring – what lay ahead around the bend in the river – until his boat passed the bend and made that part of the river the present. Thus, time on the river was really in three segments for him – the past he had left behind, the present he could see, and the future he could not see. The only factor that brought continuity to this passage of time was linkage created by the river itself. Another man was in a plane flying over the river. This man could take in at the same instant what appeared to be the past, present, and future to the man in the boat. The man in the plane could see the part of the river the boater had passed, the place in the river where the boatman was now, and the landscape ahead of the boater on the river and around the bend. The man in the plane viewed the boater's past, present, and future as a single unit. Time was perceived by the man in the plane as a single event. Which man had understood time properly, Balderston asked the listener, or have both men understood it?

L.M. Montgomery, author of the beloved book *Anne of Green Gables*, regularly dipped into the past in her journals, in order to understand the linkage of past to present. She remembered a particular summer evening, for example, at her old home in Prince Edward Island. Her grandparents were doing farm chores at dusk as she idly watched. But what struck her that evening was the frolicking of the young lambs in the pastures and the placid grazing of their dams. The vision held huge meaning for her and seemed, for some reason, to be part of the present. It had collaged past to present and resonated with precious memory in her mind. I thought of calves running and playing in the evening light in our farm's front field, while their mothers grazed quietly. The vision always made me remember evenings at my childhood home in Toronto, evenings in spring spent playing outside, climbing the cherry tree, listening to the whistling of robins while adults were indoors performing the daily tasks. Montgomery's words could resonate with me, through my experience with time as a collaged memory – the running calves to my own childhood, combined with the sound of robins at any time in the past, present, or perceived future.

If we are able to collage memory, we can also splinter it into flittering isolated events. It is surprising how powerful these memories can be. I realized that when I remembered how striking some of my own simple, short memories had been. A vision of my children, Alison and David, in bathing suits, climbing the hill in the field on a mid-summer day springs to mind. It comes back to me so clearly: the blue sky, the warm ground, the hum of flies, and the glint of sun on burnished blond, childish heads. Sun-tanned legs, backs, and arms spoke of summer. When I watched that event, suddenly I felt that everything in my life was right – it was not just happy. Everything was safe, everything was stable. Why? Yes, it aroused the sense of all of yesterday's summers merged into one. But more, it felt like the present and the past at the same time. All summers became merged into that day. Is that what *déjà vu* means – the sensation associated with the sense that time can stand still? Time standing still provokes a sense of beauty. Time standing still can also be healing.

~~~

Mammograms and mountains. As I lay on a hospital table one time, waiting for an ultrasound following a mammogram, a memory of a split second engulfed me and I felt safe. The present slipped away, and I was with my family again at Killarney. We were on the mountain. I could feel the hush of the forest and see clarity in the tumbling of the mountain stream's water as it fell over translucent, marble-like, white silica rock. The sweetness of the air echoed a silence that was punctured only by the noise of the stream as it flowed down over rocks and by the high soughing of the overarching sugar maples' branches. Light poured through the tree canopy, making delicate patterns on the bright green forest floor, so filtered by those trees that it took on a sense of carpeting. I felt protected by a sense of timelessness that the memory provoked, even before the ultrasound revealed no signs of breast cancer.

Memories of split seconds – short, crystallized images – have the ability to take on a force disproportionate to the significance of the event remembered. Such memories carry a sense of importance, can provoke nostalgia, and can even stand to represent all periods of time. Short memories sometimes tell us more about the meaning of time than do long or detailed memories. Even mere splinter memories can represent or speak for longer and extended passages of time. Is it possible that a single second can describe a lifetime?

The barn had taught me that nostalgia involves pain. But what can we make of the pain aspect of nostalgic memory? We may appreciate time passage as a collage, but when we do so it is often accompanied by a strange sensation of pain. How does this pain relate through memory to creativity and learning? Is the pain aspect of nostalgia part of its beauty? Does the motivation to write journals tell us something about the linkage between pain, creativity, and knowledge? My mother, who graduated from law school in 1933 but did not pursue the profession she was trained for, wrote somewhere that happy people do not keep journals. A curious thought, first, because I wonder if it is true, and, second, her writing the idea down might mean that it says something about her. Was she saying she could not think of something worthwhile to say or that she was not driven to say anything because she was happy? Or was she really revealing that it saddened her to know she had no compelling thoughts to record and was asking herself why? Later I noticed that L.M. Montgomery asked the same question: Why do people keep journals? Her answer was that only lonely people do so. Yet separately she admitted, in her quest for knowledge, that she found it hard to write about happiness in her journal and easy to write about pain. The content of her journals certainly confirm that difficulty. Her agony in them is palpable.

Pain, creativity, and learning – some intimate relationship seems to exist between and among these phenomena. Perhaps this idea can take us closer to an explanation for why people keep journals. It seems to me that they do it to preserve memory and, in the process, create, because they want to understand themselves. The process of learning from memory often involves a move from discomfort or pain to a sense of pleasure that has been called knowledge, beauty, truth, or love. As the great Greek tragedy writer Aeschylus said so many centuries ago: "God, whose law it is that he who learns must suffer. And even in our sleep, pain that cannot forget, falls drop by drop upon the heart, and in our own despite, against our will, comes wisdom to us by the awful grace of God."

Aeschylus was not the only great writer to perceive that an intimate connection exists between time perception, the desire for art, and the relationship of pain to both, as I learned from reading the thoughts of a number of philosophers in the collective readings of *Philosophies of Art and Beauty*. John Dewey concurred: "Art celebrates with peculiar intensity the moments in which the past reinforces the present and in which the future is a quickening of what now is." In other words, art has strength when it celebrates time as a unity. Dewey elaborated on time, pain, and art as follows. "Even when not overanxious," he explained, "we do not enjoy the present because we subordinate it to that which is absent. Because of the frequency of this abandonment of the present to the past and future, the happy periods of an experience that is now complete because it absorbs into itself memories of the past and anticipations of the future come to constitute an aesthetic ideal. Only when the past ceases to trouble, and anticipations of the future are not perturbing, is a being wholly united with his environment and therefore fully alive." And, for Dewey, that "delightful perception" forms an aesthetic experience. But he emphasized that memory was critical in this process. "A return to a scene of childhood that was left long years before floods the spot with a release of pent-up memories and hopes," he wrote. "To see, to perceive, is more than to recognize. It does not identify something present in terms of a past disconnected from it. The past is carried into the present so as to expand and deepen the content of the latter."

Time passage – seen as a collaged event, whether that be explained as unity, continuity, sameness, change, fluidity, nostalgia, beauty, truth, or love – plays a critical role in thinking, and therefore in art. The more I read I saw even better how widely that idea had been accepted by others. Marcel Proust's *In Search of Lost Time* is an extended and magnificent study of the meaning of memory. Dewey quoted George Eliot on the subject of the past teaching the present, and the past transforming perception of the present into love. Vladimir Nabokov, in *Speak, Memory*, explored the themes of time, memory, and knowledge. "I may be inordinately fond of my earliest impressions, but then I have reason to be grateful to them. They led the way to a veritable Eden of visual and tactile sensations," he wrote. Time, he noted, was inextricably interwoven with memory. "The beginning of reflexive consciousnesses in the brain of our remotest ancestor must surely have coincided with the dawning of the sense of time," Nabokov stated. He then reminisced. "A sense of security, of well-being, of summer warmth pervades my memory. That robust reality makes a ghost of the present. The mirror brims with brightness; a bumblebee has entered the room and bumps against the ceiling. Everything is as it should be, nothing will ever change, nobody will ever die."

Jose Ortega y Gasset said that, to be human, is "to be able to continue one's yesterday today without thereby ceasing to live for tomorrow; to live in the real present, since the present is only the presence of the past and future, the place where the past and the future actually exist." He elaborated in the following words:

> The future is always the lead, the *Dux*: the present and the past are always the aide-de-camp and soldiers. We live moving forward into our future, supported by the present, with the past, always faithful, off to the edge, a little sad, a little frail, as the moon, lighting a path through the night, goes

with us step by step, shedding its pale friendship on our shoulders … My future, then, makes me discover my past in order to realize that future. The past is now real because I am re-living it, and it is when I find in the past the means of realizing my future that I discover my present. And all this happens in an instant; moment by moment life swells out into the three dimensions of the true interior time. The future tosses me back toward the past, the past toward the present, and from here I go again toward the future, which throws me back to the past, and the past to another present, in a constant rotation.

Ortega extended that idea by linking contemplation of time unity to an act of love. He explained the pleasure we experience with revelation as the sensation of love.

The above passages describe with clarity and power my theory that nostalgic memory relates to art, through the sensation that time is collaged. Why, then, am I pursuing my memories by writing this book? I could have simply read what others have written, in order to find the answers I was searching for – answers to how aesthetics results from memory, and the effects of time passage when it is collaged in the mind. But would I have read the thoughts of others and valued them the same way if I, too, had not pondered some of the same issues? I don't think so. There is immense value in going down your own pathway to meet Mnemosyne and the muses through your own memories. The words of great thinkers make better sense that way, and the larger significance of the goddess and the muses is more fully understood. We do not really appreciate how we think unless we know how our past makes us interact with reality. Collaged memories of a personal past teach us about humanity collectively.

As I came to appreciate how significant time and memory were to aesthetics for many writers, the relationship of beauty and truth to either time passage or art increasingly puzzled me. I had, in my art exhibition of 1991, placed beauty with truth, and time passage with both. Impressions about the past certainly relate to art, as do beauty and the truth. Is perception of time as continuity beautiful because it is the truth? Perhaps, but that does not mean that truth is synonymous with beauty. It has been said many times that truth is beauty and beauty is truth. But is all truth beautiful? Certainly not, at least on the surface. But why can the ugly and the terrifying in art provoke sensations of perfection and, through perfection, beauty? Or can we confuse a sense of perfection with a sense of beauty? Is the sensation of perfection synonymous with beauty or with the truth? It seemed to me that the answer might be that perfection is felt when we perceive that the truth can be understood. Perfection is a revelation of truth. Art seeks revelation. Perhaps it is revelation, a form of truth, and not beauty that lies at the base of aesthetics, I thought. Beauty is one form of truth under this theory, however, and therefore is one form of art. The point is that it is not the only form and, therefore, it does not define art. Lack of concealment might, ultimately, be what art strives to achieve. In this I show my affinity to Martin Heidegger, a twentieth-century German philosopher, who saw truth as unconcealment. Benedetto Croce and Dewey also denied beauty its exalted position in aesthetics; they saw art as an expression or a symbol, not a reflection of beauty. Ortega placed love above beauty in aesthetics, by making love the seat of all aspects of philosophy. But let Heidegger speak: "Truth is the unconcealment of that which is as something that is. Truth is the truth of Being. Beauty does not occur alongside, and apart from this truth. When truth sets itself into the work, it appears. Appearance – as this being the truth in the work and as work – is beauty."

I began to suspect that all these themes – memory, transformation of memory, aesthetics and nostalgia, the meaning of art and reality – were intimately interrelated, but not through the commonality of beauty. They could actually be all facets of the same thing – an avenue to revelation, which results from the evolution of understanding through Mnemosyne and the muses. My memories and my study of various aspects of philosophy ultimately made me understand how entangled these thoughts really are. Not only is it true that my past is important to my appreciation of the present and the future but it now seems to me that impressions of the past generally wire people's minds for their outlook on both present and future.

Time and memories form various complex collages. In turn, collaged vision can get recycled and, as a result, create a deeper or layered memory. Art could be made from collaged memory because such imagery would feel like a revelation of the truth to the viewer. Mnemosyne and the muses had made these things clear to me. They could sing "the story of things of the future, and things past," and in doing so they could, as Hesiod tells us, assuage pain. They could do more, though. They could also relate science to art.

Summer Grazing,
oil on stretched canvas,
100 x 127 cm

David and Aster,
oil on stretched canvas,
127 x 100 cm

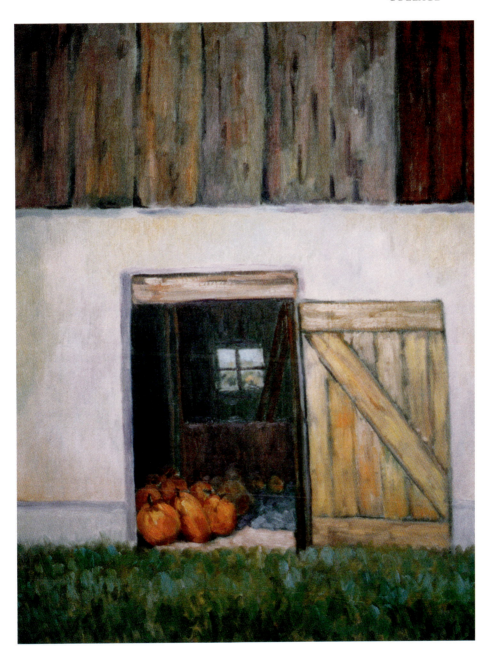

Barn Door,
oil on stretched canvas,
60 x 76 cm

Jalna and Silver,
oil on stretched canvas,
60 x 76 cm

Panda on a Hilltop,
oil on stretched canvas,   *53*
100 x 76 cm

Alison at the Rock Pool,
watercolour on paper,
38 x 55 cm

Summer Dreaming,
watercolour on paper,   *55*
55 x 38 cm

Alison,
watercolour on paper,
55 x 38 cm

David,
watercolour on paper,
55 x 38 cm

Hesiod, an ancient Greek poet, was tending his sheep on Mount Helicon almost three thousand years ago. The nine muses appeared to him and said, "We know how to speak many false things as though they were true; but we know, when we wish, to utter true things." "They breathed a voice into me," Hesiod said, "and power to sing the story of things of the future, and things past. They told me to sing the race of blessed gods everlasting, but always to sing of themselves at the beginning and end of my singing." Hesiod would tell the truth about life, would know how to do so, through the power of art and science, memory and history.

    ~ Hesiod, "Theogony"

CHAPTER THREE
Dinosaur Bones

Time passage and memory as collage – fused together and encoded in my mind – would lead me into ever greater knowledge. To begin with, I began to recognize that there were other pasts not related to me, my family, or my culture, and that these had come to play a role in my perceptions about the meaning of time. I would appreciate the phenomenon of extinction through a suffusion of childhood memories into a growing intellectual desire to understand how time passage affects the very essence of existence itself. Quite simply, I became interested in evolutionary biology because of childhood memories.

I was not alone in these sensations. The fascination many people have for the process of extinction (and through it, evolutionary biology) often begins in childhood and is entangled with personal memory. Dinosaurs are particularly important to this process. Stephen Jay Gould, for example, said he owed his lifelong concern with paleontology to his visit, at the age of five, to the Museum of Natural History, where he saw the skeleton of *Tyrannosaurus rex*. But dinosaurs and paleontology would lead him into his true love, the study of evolution itself. "For some reason that is still unclear to me," Gould wrote towards the end of his life, "I always found the theory of how evolution works more fascinating than the realized pageant of its paleontological results, and my major interest therefore always focused upon principles of macroevolution." Evolution is the study of time's dynamics. Paleontology is the study of time's effects. Both are studies of time.

Dinosaurs began to interest me too at the age of five, when my mother and father took me to see *Fantasia*, my first movie. There I watched *Tyrannosaurus rex* battle with *Stegosaurus*, as rain pelted down on the creatures. The scene left complex and vivid sensations in my mind, which did not fade after learning in adulthood that *Stegosaurus* had been extinct for millions of years by the time *T. rex* was around in North America. In the movie I also saw dinosaurs walk to extinction, over hot sand, leaving only their bones, and I would imagine sand drifting over those skulls forever in my mind's eye. The vision always aroused pity in me, even though here, too, scientific truth, as I learned later, was missing. *Stegosaurus*, a dinosaur that died out early, plodded on beside *Parasaurolophus*, a species that existed late in the last dinosaur period, the Cretaceous. The unending relationship of life to fighting and death, and the subsequent sense of fear that these patterns arouse in the human mind, ultimately became the real messages that never left me. Even as a small child I recognized I was seeing basic truths about the reality of existence. I was, in the process, also experiencing something overarching about the impact of time on thinking.

Within a year or so of seeing *Fantasia*, my father took me to see dinosaur skeletons at the Royal Ontario Museum in Toronto. A mounted *Lambeosaurus* skeleton here moved me the most. Its dimensions were truly awe inspiring. I gazed, over and over again, at its towering height, at its massive feet and hind legs, its small hanging forearms, and its tiny head that stood so high it almost seemed to touch the vaulted ceiling. Over childhood years, I went back many times to the museum to look at dinosaur bones. I begged my father to take me there on Sunday afternoons. By the time I was an adult, the process of extinction, the study of time's effects on existence, fascinated me. I read as much as I could on paleontology. I returned to the museum many times to look at dinosaur skeletons and to read dinosaur literature in the museum's library. Drawings of bones in scientific journals I found beautiful. I went to New York, where I, too, saw a full skeleton of *Tyrannosaurus rex*. I visited the Tyrrell Museum in Alberta and gazed over the Badlands, where so many Cretaceous dinosaurs were found at the turn of the twentieth century.

Dinosaurs personified something about time for me, but in a layered way. Their transformed bones linked the past to the present in an empirical sense: even though extinct, their physical presence remained in stone bones. Dinosaurs were also about my past, through memories, and about my present as well, because I could see the bones today. Instinctively drawn to them because of their relationship to the meaning of time, I would finally come to realize that they impinged on me in a double way: aesthetically and scientifically. Dinosaurs could be art. Dinosaurs presented myriad ideas that were aesthetic for me, through nostalgic memory. Drawing bones from life in the museum made me feel a sense of beauty, or symmetry and balance; and I found I could use colour and light to distill nostalgia. The more I drew the bones, the deeper my interest in paleontology became. Reading about that subject drew me into a study of changes in anatomy over time and within different species. I learned about mechanics, because mechanics explained how bone and muscle work together. I was, in effect, approaching the problem of evolution itself, but from a multifaceted point of view. The underlying thought that I was relating art to science and to time, through memory and a focus on a historical science, aroused an awareness in me that felt like revelation. I decided to undertake a more organized art project based on my many-sided study of dinosaur bones. I hoped to explain, through a visual medium, the fusion of art and science.

The project required carefully planned research, involving specific tasks that were both scientifically and artistically oriented. I reviewed what I had read about comparative vertebrate anatomy, and then studied the anatomy of certain dinosaur species. I realized it was important to be aware of the directions that paleontology had taken over the years with regard to dinosaur physiology. The *Lambeosaurus* skeleton I had stood in awe of as a child had, by modern standards, for example, been mounted wrongly. The extinct beast did not, scientists now believe, stand at such a height on its two powerful hind legs. The skeleton had been remounted, I found in the museum, to reflect newer theories, which are based on the assumption that *Lambeosaurus*'s pelvic could not have sustained the animal's weight at the angle needed for bipedal movement. After reading the original scientific journals that reported on specific skeletons, I understood why the bones of certain individual animals looked as they did. Injuries in life often left strange scars, and in the process brought home to me the fact that I stood in front of individual animals that once were alive.

I studied dinosaur art, and I soon saw that the beautiful drawings of stone bones that I so admired represented something different from what I was after. Executed to show exact anatomy, these drawings were designed to explain what the bones looked like, in the same way that a photograph would, for someone who had never laid eyes on them. An exhibition of dinosaur art at the museum a number of years earlier had fascinated me as well, and luckily I still had the catalogues for the show. Most of the paintings exhibited at that show depicted what dinosaurs might have looked like in life, and therefore did not focus on my problem – the fusion of art and science as a theme. I was after something different – and the only way to test my vision was to practise drawing and painting images of the bones in various artistic mediums. Dinosaur bones of five hadrosaurs and one ceratopid – all from the Cretaceous period, which ended 65 million years ago, and found in Alberta, Canada – became the subject matter for my theme: the fusion of science and art.

The next part of my research entailed the drawing and painting of various studies. I spent many hours in the Royal Ontario Museum, drawing bones from life in an effort to explore colour, light, and form. I would sit against the wall and, using coloured pencils, which allow for the layering of pure hue, I would explore what was in front of me. I found I could use delicate shifts in colour to express what the vision felt like. A pink-violet, for example, suggested how the softness of a reflected hazy shadow, cast by a distant light that perhaps was meant to illuminate something else on display, kissed a bone accidentally and beautifully. Executing black and white drawings, done from life as well, allowed me to investigate the problem of composition. I focused not only on bone pieces but whole skeletons as well. Next I did a series of watercolour paintings in my studio to test the strength of artistic content found in the many drawings. This exercise made me reshape the original compositions found in the life studies. I was then ready to experiment in oils. Because I wanted many of the bones to be presented in life size, I had to see how well the newly planned compositions would work when greatly increased in size. I assessed values – light in relation to dark – in these sketch canvases as well.

Next I executed twelve oil paintings mostly on stretched canvases that I had built myself. For me, the canvas is part of the creative act, and I love the feel of the heavy material as I stretch it across the wooden frames.

I began with my old favourite, *Lambeosaurus*. In *Complete "Lambeosaurus lambei" from Behind* I wanted to show that it was possible to say something about size, light, and colour with paint in a small, simple way. Images that do not dwell on details can send out powerful messages. This painting is a greatly simplified version – with respect to anatomy representation, painting technique, and physical size – of the newly mounted *Lambeosaurus* skeleton. Exactitude of depiction, with respect to anatomical correctness, did not increase my sense of the beast's monumental size. As I looked over my drawings, I saw that *Lambeosaurus* offered me another but completely different compelling image, which explained aspects of

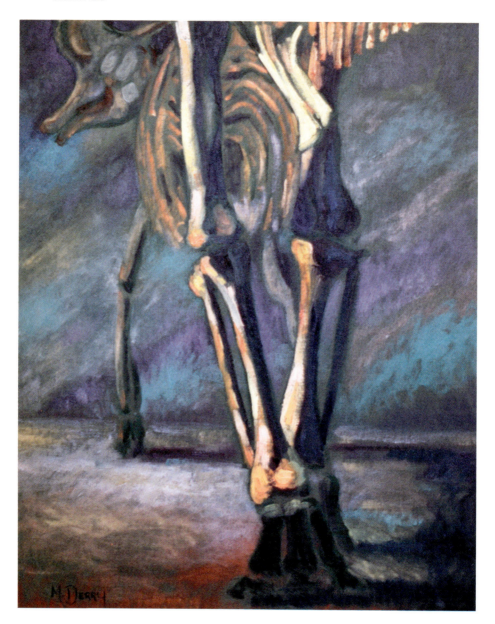

Complete *"Lambeosaurus lambei"*
from Behind,
oil on canvas board,
76 x 60 cm

"Lambeosaurus lambei"
Skull Straight On,
oil on stretched canvas,
100 x 76 cm

dinosaurian life. Because the skeleton was now mounted as walking on four legs, I could look more closely into the face of my old friend. In *"Lambeosaursus lambei" Skull, Straight On*, I confronted what is for me the strangest image of a dinosaur. The narrow skull, mounted on the cervicals, confused me, and I instinctively looked for the eye sockets. The orbits of the skull lie well back and are not the dark upper grooves of the crest. I was suddenly convinced of the probable poverty of the braincase. The natural strangeness of the skull lent itself to abstraction, I decided, or to non-literal representation. Abstraction is enhanced by the softened patterns, spilling luminous light, and sweet colour that breathes through the atmospheric "background." I used luminous colour to imbue this painting with dreamy mellowness and to create great space behind the skull. In doing so, I felt a sense of time suspension in relation to, or perhaps I should say contradicted by, the rigidity expressed in the frozen stance of the skull and cervicals.

The huge skull of the one *ceratopid* I studied aroused different sensations in me. In *Through the Skull of "Chasmosaurus,"* I found myself more interested again in united shapes than in the literal forms that described the complex bone structures, yet I wanted to say something different about space. This painting was designed to repress a sense of depth. I used colour and light to simplify and flatten the forms, and to reduce a feel for distance and volume in the background. An overall mosaic of colour was what I was after in this portrayal of anatomy – bones working together, as they do in life. For me, this dinosaur carried a sense of the grotesque, the unbelievable, and aroused primordial fear. While my paintings of *Lambeosaurus* expressed strangeness, this one of *Chasmosaurus* said something about fear. I thought of *T. rex* and his battle with *Stegosaurus*.

The incomplete skeleton of *Parasaurolophus walkeri* fascinated me, as it has many others over the years. It is probably one of the most valuable dinosaurs in the museum. It certainly must be one of the rarest, for it is the most complete skeleton of this species ever to be discovered. I executed four canvases based on the bones of this unusual and beautiful dinosaur. All portrayed only sections of the skeleton. In *"Parasaurolophus walkeri", Skull from Behind* I wanted to say something about motion. I combined elements of colour perception (hot colours are closer, cool colours are farther away), use of light (a focus on certain bone shapes), design of negative space (strokes that echo the crest), and specific representation of form (viewed from behind the skull) to emphasize the crest. I also painted the close-to-life-sized skull with shapes designed to unify bones, and I tried to set the skull and bones in opposition to the negative space behind. The crest was intended to dominate the painting, and rightly so, for this is the most spectacular crest of the *hadrosaurs*. This dinosaur made me sense it as an individual living thing. Perhaps that was why I felt compelled to portray the interior of the skull, suggesting thinking and feeling. I wondered what happened to this particular animal and how it lived.

The theme of motion, this time through a sense of bones as machinery, also interested me when I painted *Parasaurolophus's* scapula. In *Front of the Scapula of "Parasaurolophus walkeri,"* the complex forms of the articular end of a shoulder girdle were greatly simplified, to make the bones present an antithesis to the other main subject of the work: space. I played the solidity of the bone forms against the perception of space and depth. The contrast of weight to space, I hoped, would arouse complex perceptions in the viewer. I wanted the contrast to portray suspension, because suspension could explain something about how bones work – how bones operate as machinery. Their very stillness and weight within space suggested to me their capacity for productive movement. I used sweetness of colour, here closer to the primary hues, combined with the design resulting from the way the canvas was divided up, to create a still life.

While I found all of *Parasaurolophus's* bones beautiful, I liked a particular composition, oddly enough, because of its awkwardness. Flatness and its antithesis, volume, when combined with a rather uninteresting design, seemed to say something about the nature of death, extinction, and the changes that time wrought on the bones.

Through the Skull of "Chasmosaurus",
oil on stretched canvas,
76 x 100 cm

"Parasaurolophus walkeri",
Skull from Behind,
oil on stretched canvas,
100 x 127 cm

In *Scapula and Ribs of "Parasaurolophus"* I presented flattened bone images in order to make basic simple shapes provide a design, but at the same time I maintained a delicate sense of volume – enough to allow some feel for form. Negative areas that were completely flat, though, played an important role in overall colour composition and intellectual thrust. Flatness is itself related to the meaning of death. I sensed death to the animal that possessed the rib cage, at the same time that I perceived that some force representing continuity in time made the bones become flattened and compressed after death. I was aware of death and life – both evident in the bones themselves and in their compression as well – in the somewhat awkward nature of the subject matter.

The Pelvis and Femur of "Parasaurolophus walkeri" is also about the nature of death and extinction. Portrayed just under life size, I used these bones to describe anatomy and perception of light. Emphasis on light results from the contrast of dark to light. The femur hangs heavy, dark, and huge against the highly coloured, lighter background. The image was intended to carry the sense that death and life, past and present, relate to each other in a critical way. Death, one side, is portrayed as extinction and is linked to dismemberment (evident in the missing tibia and fibula). Life, the other side, is intertwined with death because the painting describes the present, through its emphasis on light and colour perception as equal subject matter. The angle of the femur suggests movement and stillness at the same time: life and death.

The Skull of "Anatosaurus edmontoni" says something more about the meaning behind stillness. Here I saw stillness as a state dominated by frozenness. The rich, rough action of the paint was intended to suggest a movement that is, in turn, denied by the lack of depth in the painting. The warm and bright hues, found in the coloured negative space on the canvas, bring the area around the skull forward. The lack of depth is accurate, for the bones are embedded in the matrix stone. The skull is held, so to speak, still. Even so, and in spite of the flatness, I hoped that form and volume could be read into the bone shapes. This dinosaur felt the most reptilian to me, and my mind easily linked the animal to present-day living reptiles. Continuity with extinction.

Perhaps nothing says more about the monumental size of these beasts than *Pes of "Anatosaurus edmontoni."* The bones here are nearly life size and, because they represent such a small portion of the body, the painting carries a sense of the animal's hugeness. The painting's use of colour was designed to make the viewer feel the presence of beauty detached from the subject matter. Colour appreciation results from the way we understand light from an aesthetic point of view. I mean the way we feel, not think, about our perceptual interaction with light. There is something bird-like about this gigantic foot – and with good reason. Today we believe that dinosaurs were the ancestors of modern birds.

The Skull of "Kritosaurus incurimanus", as with the image of the *Anatosaurus* skull, is a painting in some anatomical detail of a skull embedded in the matrix. Here, too, a sense of flatness and stillness pervades. But the stillness does not have a frozen quality for me. I was far more aware that the stillness was graceful and, as a result, I found this to be the most beautiful stance of the dinosaurs found in the Royal Ontario Museum. The arching gesture, created by the position of the cervicals and skull together, is profoundly moving: an action, frozen in time, simultaneously suggesting to me the tortures of death and the movement of ballet. I chose a painting technique that was designed to support that twin sense. The circles in the negative areas formed by the matrix repeat the gesture to portray spilling light, and also to provide the dual perception of flat space and design abstraction. I suppressed the natural light/dark of the bones resulting from lighting so as to give greater unity to them, and to enhance their integrity with the circle shapes.

In my second image of this dinosaur I explored colour somewhat differently from the way I had in any of the other paintings. In *The Pelvis of "Kritosaurus incurimanus"* the bones were allowed to take on an abstract design through the representation of natural light/dark

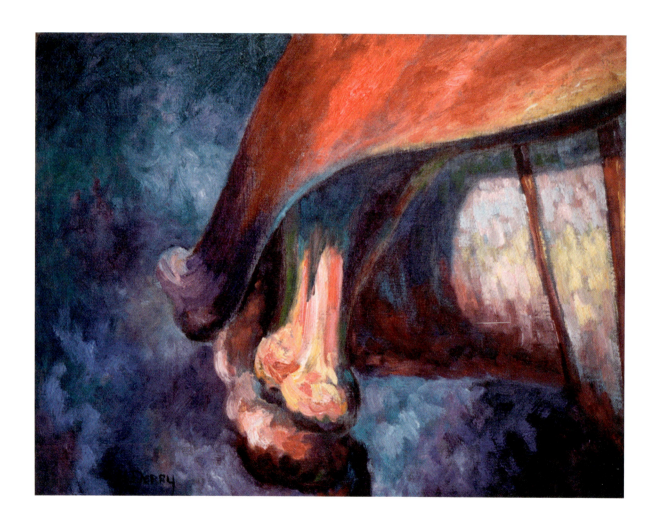

Front of the Scapula of
"Parasaurolophus walkeri",
oil on stretched canvas,
40 x 50 cm

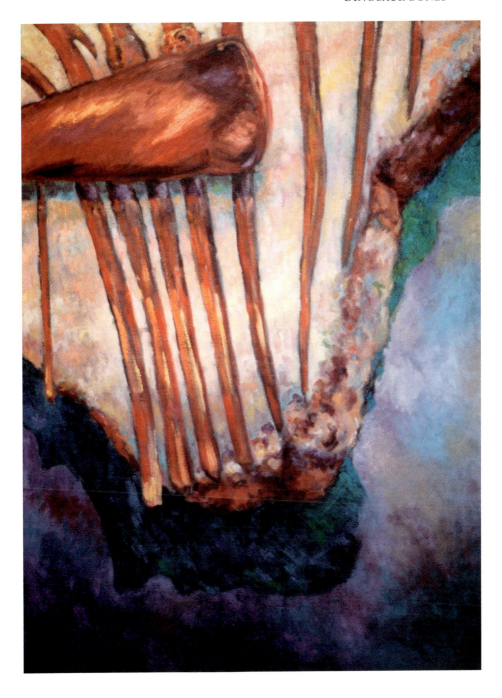

Scapula and Ribs of
"Parasaurolophus walkeri",
oil on stretched canvas,
100 x 76 cm

73

The Pelvis and Femur of
"Parasaurolophus walkeri",
oil on stretched canvas,
127 x 100 cm

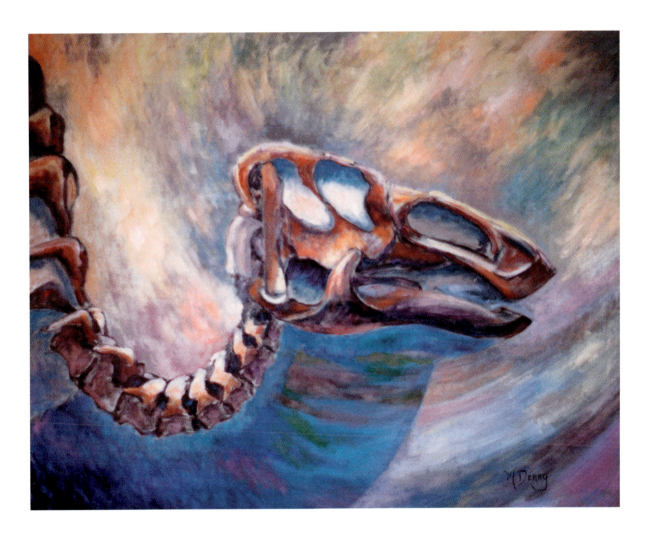

The Skull of
"Anatosaurus edmontoni",
oil on stretched canvas,
100 x 127 cm

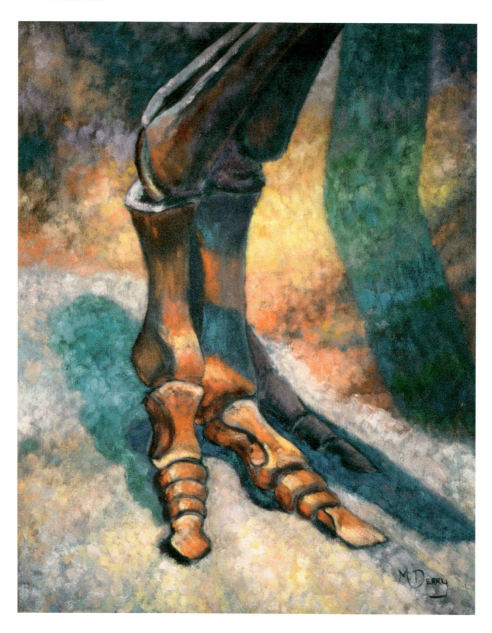

Pes of "Anatosaurus edmontoni",
oil on stretched canvas,
76 x 60 cm

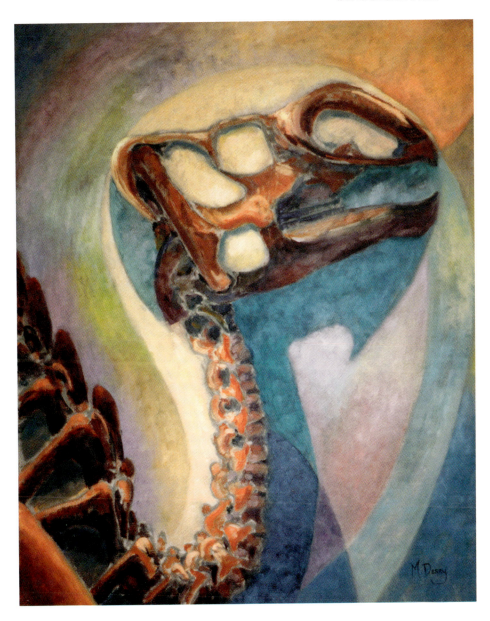

The Skull of
"Kritosaurus incurimanus",
oil on stretched canvas,
127 x 100 cm

The Pelvis of "Kritosaurus incurimanus",
oil on stretched canvas,
76 x 100 cm

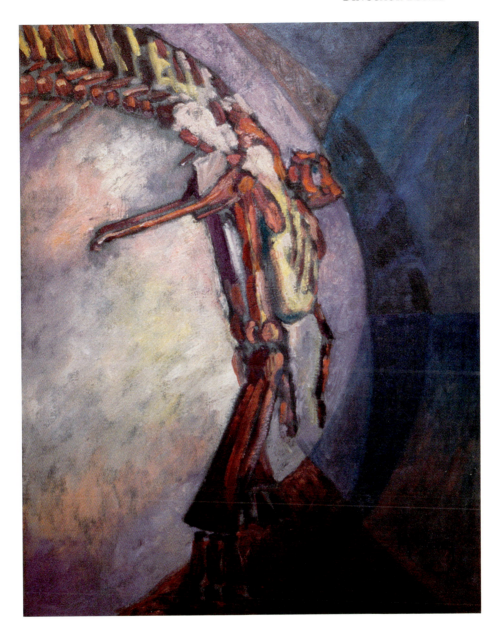

Study of "Prosaurolophus maximus",
oil on stretched canvas,
50 x 40 cm

patterns that resulted from the lighting – spilling overhead light. Brilliance of colour was repressed in order to give more force to the representation of shapes. Colour varied more by light/dark than by hue. As a result, this work is the most achromatic of the paintings. The general flatness of space, caused by the matrix, was intended to enhance the striking patterns of paint. Form was presented, however, to provide some sense of volume. I wanted the various naturally abstract shapes in the matrix and in the trellis of tendons to add areas of interest to the painting.

The small *Study of "Prosaurolophus maximus"* was sheer fun to paint. I played down anatomical correctness of bones in this image. Perspective exaggerated the flow of the tailbones into the vertebrae of the back, and, in doing so, suggested a circle shape. I used that suggestion. Circles symbolized symmetry, and I made circle shapes dominate this painting – literally providing the design for all negative space. I intended the circles, due to the sweetness of the colour, to suggest air bubbles and/or lenses revealing a dinosaur. I wanted to present dual symbolism: ephemeral bubbles, or fragility, and telescopic lenses into the past, or lastingness.

~~~

I looked at some of the talents of the muses, but within the context of memory, when I undertook this project. I painted my feeling for extinction. My interest in the subject began long ago in my distant past and came to full flower in adult years. Mature interest developed, however, within the framework of the original fertile ground that had been laid down in childhood. Each image was designed to express something in particular, and each contributed to seeing two aspects working together: the art in the paintings and the science of paleontology. The twelve canvases portrayed vertebrate anatomy of certain extinct species, but they also presented my thoughts on the way we appreciate colour and light. The collection described various ideas – some philosophic, some artistic, some scientific – through a medium that is flexible and elastic enough to carry infinite variations of complexity. There were many more interwoven meanings in the art for me than I have captured here. It seems I don't need to elaborate on those meanings. They come and go through my mind, and that is as it should be. I like the way the art pieces can call forth different splinters from aspects of an understood, and sometimes not understood, past in me. The language of images is often more capable of expression than the language of words, whether applied to aesthetic or scientific ideas.

Understanding extinction on various levels – or exploring the meaning of extinction within the framework of our lives – had, through acts of painting, led me into an important overarching question. How is aesthetic judgment related to scientific judgment in all the ways we think? It had seemed to me that I perceived everything around me artistically (or with an appreciation for beauty) and empirically (or "scientifically"). My painting and my study of dinosaur bones in the end made me want to understand reality separately from memory. Dinosaur bones, in effect, launched me onto a new part of my pathway.

The gods revealed to people "the rising of the stars and settings hard to judge. And then [the gods] found for [people] the art of using numbers, that master science, and arrangement of letters, and a discursive memory, a skill to be mother of Muses"

   ~ Aeschylus in "Prometheus Bound"

CHAPTER FOUR
Light and Shadow

Painting dinosaur bones allowed me to see that by collaging memories, I could do much more than access an aesthetic sense. I could increase my basic ability to learn. Memory and nostalgia had encoded me with a way to think, and had, in the process, become the underpinnings of a deepening desire to acquire knowledge outside myself. I started, for example, to see the physical world around me with new eyes. It thrilled me, and I spent a lot of time simply looking at its boundless beauty. Sometimes I tried in words to capture aspects of its essence, so that I could project that reality in paintings. I saw the way light and shadow shaped vision in everything I looked at: the meaning of light and shadow for visual perception came to fascinate me.

I watched their effects on snow as I tramped across fields on brilliant sunny winter afternoons, and I analyzed the light and shadow in terms of the three characteristics of colour they presented: their chroma, or brightness of saturation; their hue, or what we commonly think of as colour – greenness or blueness, etc.; and, the most significant of all, their value, or light to dark, in relation to the other two qualities. I saw that we can relate the chroma of the sky to that of the shadows on the snowy ground, but that the sky's blueness, or hue, is a warmer, yellower blue when compared to that of shadows. These are a colder and darker blue, but just as luminous, if not as hue saturated. As the day advances, shadows on snow change hue and take on a violet cast. They darken in value as well. Sunlit patches become brilliant yellow by that time. Reddish light from the sun at the end of the winter's day creates an orange-red on the sunlit tree trunks. As the shadows lengthen, and they do so dramatically late in the day, the sunlit areas of snow take on a pinkish-vermillion cast. The same rosiness can be seen in the clouds down sun. The wind-carved shapes of snow, like small waves or ripples, smooth or sharp, on a lake reveal their form dramatically in the low sun. Tints of pink-violet are also seeable in sunlit areas of snow – dancing with luminous colour.

If a winter day is beautiful through light and shadow, I soon realized that a winter night in the country can be even more so, under illumination from moon and stars. I often went out in the dark on winter nights when we were at the farm on weekends to look around me and try to understand what I saw. Moonlight playing on snow creates a breathtaking world that is completely different from that made in sunlight. The sky is dark, but studded with stars. The snow does not glitter to the same extent, I noticed, but lies flat and dreamlike, with a suppressed shimmer, in the strange light. The farm house's stone walls can be seen clearly, and even the hue of the stone is perceivable – warm and pinkish – quite startling to see in a pre-dominantly achromatic, or black and white, world. Icicles hanging from the roof and directly facing the moon glisten and glitter in that light. The greatest darks are, of course, the naked trees. It is curious how closely the values of the moonlit sky are to the snowy ground. Shadows act as mid-tones between the darks of the trees and the light of the snow and sky. Cast shadows are vital to comprehension of a moonlit scene. They confirm the sense of light from the moon and stars. Without shadows, the dark of the night allows for no sense of form, because in darkness we cannot perceive form. It is by value, primarily, that we see volume. And we need shadows, just as much as we need light, to understand the relativity of values. Illumination by moonlight seems strong, but actually is weak enough that the use of peripheral vision often allows for better perception.

I looked forward to seeing how moonlight affected a spring landscape at night because the ground's chroma, hue, and value in that season are very different from those in winter. On a May night in the full moon, I walked over the lawn at the farm, gazing at the fields and down the lane through the trees. All around me was the music of frogs, a penetrating chorus rich with the promise of spring. And I did find interesting variations with the coming of spring. While the sky was lighter than the ground, the ground was not as dark as I would have expected. The trees cast shadows on the grass, just as they do on the snow. Trees and woods were dark and dense. They seemed impenetrable. Form in the trees was hard to distinguish; they presented a lit side and a dark side, but little sense of roundness. More distant tree shadows displayed reduced value – true to a sense of depth. Sides of the birches, facing the moonlight, were extremely luminous. White objects everywhere – barn foundations, for example – gleamed with brightness. Grass showed texture at close range. Where the sky met the ground in the distance, at the top of the field and at the end of the lane, demarcation between them became blurred. Sky and ground blended into each other.

Light and shadow together present a sort of dichotomy, essential to visual understanding. Dichotomous perception dominates all human thinking, as Stephen Jay Gould has pointed out. Gould argued that we need to understand every idea, every thought in terms of contrast through pairs, or within the framework of opposites. Perceptions of light and shadow are examples of the dichotomous approach to reality that is encoded in our humanity. Perhaps light and shadow perception is more than an example. Perhaps it explains why we tend to think generally in dichotomous terms about abstract thought. We are not encoded to do anything else.

Observing the effects of light and shadow made me see how much we needed both, in order to comprehend the world. By looking at the beauty, symmetry, and balance that could be found in nature, through the effects of light and shadow on vision, I was also trying to appreciate the force of natural laws – their governance of the world around me. I wondered, why could shadows look so different? Why were some sharp and some blurred? I noted that shadows of trees were blurred when the surface where the shadow fell was farther away from the tree. I looked at the tree shadows on the garage in the early morning sun. The maples overarching the building cast shadows that were sharp and dark on the roof. Against the garage's side, however, fell soft and blurred shadows of tree trunks, cast by the more distant line of pines. I noted as well that long shadows, cast by trees in low or late-day light, tend to be more blurred and soft than those that were short and cast by trees in high or noonday sun. I was seeing the effects of light in relation to both the object that casts the shadow and the shadowed surface. I also observed how aspects of vision worked. When I looked at something through a frame of my hands, I found it no longer looked the same. We see in a semi-sphere and are influenced by visual frames that lie beyond what we are actually looking at. I became increasingly aware that I "saw" with all my senses, and I wanted to portray what my eyes were incapable of revealing to me. I wished I could paint the sound of the wind.

I was trying to understand the effects of light and shadow in both "artistic" and "scientific" terms. I was attempting to see all reality the way I had the dinosaur bones. Was science essential to the production of all art, then? In what way was I actually thinking, when I created any piece of art – art that might have nothing to do with dinosaurs? I assumed I thought aesthetically, but what did that mean? Maybe more knowledge about aesthetics itself, outside the way memory and time had affected my sense of art, would make me understand better the processes of my thinking when I was in the act of painting. I set out on a new voyage of discovery, in relation to my painting. My technical ability had steadily improved with my stumbling attempts to learn about art, but I was not so sure about my vision – the heart that lay behind any image I executed. Had it matched that same type of growth? And if it had not, how could I expect the fundamental quality of my work to get any better?

I looked at the writings of Immanuel Kant, Friedrich Wilhelm Schelling, Georg Wilhelm Hegel, and Arthur Schopenhauer in relation to aesthetics. In other words, I began to read about the philosophy of art. It was then that I learned just how significant was the theme I had tried to paint in my dinosaur paintings. I was confronting a major underlying problem, from a philosopher's point of view, in all human thinking: What is art's relationship to science? But what about the actual production of art, in relation to artistic and scientific thinking? The metaphysical approach of these philosophers did not clarify for me the process of making art. I did not know more about what made me paint or how artistic and scientific thinking went into my acts of painting. Abandoning the academic exercise of reading the comparative works on aesthetics, I focused instead on the work of two people who seemed to explain the art process itself: Joshua Reynolds, a painter, and Friedrich Nietzsche, a philosopher. Both were less preoccupied with metaphysics than the former writers, and the wonder that art aroused in the human mind underlay their thinking. They seemed to say more about the mechanics of the aesthetic spirit, explaining better the act of painting. I thought this approach would also show me how art related to science.

Reynolds's noble descriptions (collected in *Discourses on Art*) of a painter's ideals, which that great artist presented over many years to the Royal Academy in London, inspired me. What Reynolds said was based on classic Enlightenment thought: he believed nature could be understood outside the self. "The great business of study is, to form a mind, adapted and adequate to all things and all occasions; to which all nature is laid open, and which may be said to possess the key to her inexhaustible riches." He may have spoken about beauty, but his approach to art was "scientific" when compared to Nietzsche.

Nietzsche seemed to dissolve the art/science conflict in his description of art. He held that any interpretation of reality rested on art because existence could only be understood through aesthetics. Science, for Nietzsche, was subservient to art. But art, or the contemplation of art, produced a delicious agitation in the individual. "For art to exist," wrote Nietzsche, "for any sort of aesthetic activity or perception to exist, a certain physiological precondition is indispensable: intoxication … No art results before that happens." For Nietzsche, the human mind was always the medium to reality. This concept, of course, was at the heart of Romantic thinking: it irrevocably tied any understanding of nature outside the self to constructions within the mind.

At first, for me, art and science seemed irredeemably interlinked through these different stances of Reynolds and Nietzsche, and both approaches served as an inspiration for me to create art. But I soon saw that Nietzsche's thoughts actually contradicted my experience in painting. Nietzsche might have thought that all knowledge arose through art, but I continued to find that my acts of painting consistently indicated otherwise. As I attempted to translate my visions of nature into art, I was clearly bringing both an artistic and a scientific approach to my task. And here I was forced to think through, at least in some rough fashion, what I meant by artistic and scientific thinking. I pondered what I had read about aesthetics and what I had learned through experience in painting. Artistic thinking, for me, arose from a striving for balance, for unity, and for the incorporation of wholeness in visual language – using both the external and the internal senses together, and attempting to feel today, yesterday, and tomorrow as one. Aesthetic thinking also tends to be subjective. My paintings of children at Killarney, for example, involved a subjective approach (my memories) to project what another child could experience in any summer at Georgian Bay. Scientific thinking, for me, meant objective thinking. It resulted from empirical reasoning that arose in relation to observations made outside the self, as Reynolds put it. It strove for an appreciation of natural forces outside the self. When painting images of my children, for example, I not only focused on my internal memories but observed the anatomy of the child in front of me. I saw the child as something outside of me. The child presented a three-dimensional form and did so by virtue of light from the sun, which made his or her body cast shadows.

The problem of art's relationship to science really fascinated me now, from an academic, or a philosophic, point of view. I had moved onto another part of my pathway. I wanted to know more, even if the new information would have no effect on my painting. The conflict itself, artistic against scientific thinking, also appeared to be part of the puzzle, and it seemed to me that understanding the disagreement better might help explain how the two avenues of perception worked in relation to each other. I felt I needed to know how metaphysical ideas had changed over time, in order to gain a historical perspective.

I wandered through some articles written by Isaiah Berlin and collected in *The Roots of Romanticism*, initially because I was interested in what he had to say about Romanticism and the effects of the Romantic Movement on modern-day thinking. But I soon discovered that Romanticism was part of the larger story of art's relationship to science. As I read more of what Berlin had written, I understood how crucial one point he made was: that a fundamental conflict over how to perceive all knowledge was triggered by Enlightenment thought in eighteenth-century France. The deep divide between science and art (and between the underlying thinking in either) that seems so apparent today arose with the Enlightenment and its overriding concern with natural laws.

Before the eighteenth century, Berlin argued, the idea that science was separate from art had been much less pronounced. While Isaac Newton's laws of physics inspired the new fascination with science, the sense that all knowledge could only be had through an appreciation of principles or laws became an obsession. Reason alone could lead to understanding, and that understanding would, in turn, provide freedom and happiness. A passion for the scientific approach could affect the outlook of artists, as Reynolds's words demonstrate. But overriding devotion to reason ultimately provoked a reaction. German scholars came to despise French thinking, not only because of its intense and dry fixation with logic but because it was part of a larger French domination of Europe. Science and rationality were French; feeling and knowledge arising from the inner self were German. Philosophers in Germany developed these thoughts and came to the conclusion that science, rather than illuminating, actually acted as a prison – scientism distorts reality by setting up systems, which deny the complexity of nature. It was a reaction to the French Enlightenment that stimulated the flowering of the German philosophy of art.

I grasped a continuum, which at least explained why eighteenth- and nineteenth-century philosophic thought on art and science had evolved the way it did. I knew better how certain theories of metaphysics arose, but I was not really much closer to seeing the way art and science before that time must have worked with each other in human minds – and, as far as I knew, probably still did operate, even if now in a hidden fashion. The cleavage set up by Enlightenment thinking and its "other" – Romantic thinking – seemed to be so deeply engrained in our present-day cultural thinking that the meeting of art with science has become a foreign thought to many of us.

Modern popular perceptions concerning mathematics' relationship to the arts illustrated for me the somewhat new and strange divide. Mathematics is irrevocably linked with science in our minds today because it is an essential tool of science. This idea springs from the philosopher Immanuel Kant, who held that mathematics defined science. Science expressed itself in natural laws that worked mathematically. Kant's linkage of mathematics to science late in the eighteenth century would have serious ramifications for all endeavours of the human mind – not just the arts and sciences but the humanities as well. I will look at this idea later, particularly in relation to the nature of history. Philosophers of science and mathematics might disagree about whether mathematics is science or a separate entity closely affiliated with science, but no one seems to dispute the idea that the natural environment of mathematics, whatever the exact connection might be, is within the realm of science.

I began to think more about mathematics. The idea that mathematics might have an intimate connection with art is not an easy one to digest today, but it was not always so. Plato placed mathematics at the seat of art. He spoke of measure or symmetry as being divine and therefore basic to beauty, which by his definition meant art. Mathematics is in fact embedded in all art, and it continues to play a role in aesthetic production because it projects a powerful sense of balance and symmetry. In music, for example, the length and force of sounds, which are types of numerical measure, affect the coherence, or balance, of pieces. Mathematics is equally critical to the visual arts, as I had learned directly from experience.

When I made my earliest attempts to create art, I recognized that geometry played a fundamental role in my acts of drawing or painting. Before I was five years old or had taken the slightest interest in printing, I loved drawing and discovered, in the process, the power that perspective has in images. I did not understand what perspective meant, but I was fully aware of its effects in art. I was fascinated by the fact that I could make an image look as if it had depth and form, or three dimensions, on what was really a flat, or two-dimensional, surface – a piece of paper. I drew baby carriages over and over again, exploring this idea by trying to show that the body of the carriage had depth. I aligned the four wheels in my drawings within the framework of a parallelogram, in order to reflect the viewer's stance or angle of vision. Different parallelograms (shapes I appreciated even if I did not know the proper name for them at the time) could be made to represent the interior of the carriage, also in relation to the viewer's angle of vision. I knew these parallelograms had to match each other to make visual sense, even if I didn't understand why. By changing these geometric shapes and keeping them consistent with each other, I also saw that I could make endless variations in the images of baby carriages. As I sat in the sun on the back porch of my childhood home, with paper and crayons on my knees, I was experiencing first hand a phenomenon with a long tradition in Western art. Perspective has been crucial to Western painting for centuries. The ability to make the illusion of three dimensions on a two-dimensional surface through perspective was well understood by the early fifteenth century in Florence.

Different hues, I learned soon after experiencing the effects of perspective, have different force in coloured drawings or paintings, and the geometric break-up of a canvas surface must be planned with that fact in mind. A red spot and, even more, a glowing orange-gold one will not strike the viewer the same way that a similarly sized blue spot will, and a painter must understand that fact in order to control the relations of colours to each other. The warmth of red-orange makes it appear to come forward, bringing sensations of energy and movement, and it can arouse strong emotions like love. The coolness of blue makes it seem to recede, and causes us to feel distance, mist, or spiritual depth. The hue can trigger a sense of peace. Colour shapes on a canvas must reflect knowledge of these varying strengths and evocative sensations as well, if the image is to make sense.

Later in my painting career I learned how another geometric principle that had been revered since Greek times, the golden section, could provide measure or symmetry in visual images. In order to use the golden section in composing a painting, the larger area/length section in the art piece must relate to the whole in the same way that the smaller area/length section related to the larger section. I made the geometric principle the subject matter for the painting *Panda on a Hilltop*. The sections of sky and ground on the canvas relate to each other and to the whole as a golden section, as does the space on either side of the calf. Measure and geometry are always mathematics. Since both are embedded in art, mathematics must play a role in art. The fact that artists do not seem to revere the effects of mathematics the way their counterparts did before the nineteenth century does not change the reality that mathematics is central to art. Whatever mathematics is, the simple idea that it is not analogous to science alone was an important one. Because art and science both use mathematics, some critical link between art and science seems to exist through mathematics.

~~~

I decided I had to see the presence of science and aesthetics in actual art pieces if I was to understand the dynamics of art and science working together. Martin Kemp's analysis of images proved to be helpful, especially those that appeared in *Nature* magazine between 1997 and 1999 and were published collectively in 2000 as a book, *Visualizations: The Nature Book of Art and Science*. "My interest was less looking at the *influence* of science on art, or even visa versa," Kemp wrote in the preface of the book, "but at shared motifs in the imaginative worlds of artist and scientist." Would this approach show how art and science worked together in thinking? I wondered.

The specificity of Kemp's thinking, when seen in relation to what he concluded more generally, struck me as being especially valuable. He analyzed particular pieces of art to illustrate his thoughts on art and science, ideas he hoped would reveal overarching patterns. Through paintings, he explored the idea, for example, that perceptions which drive art movements can be triggered by science. He noted that the Cubists, especially Picasso and Braque, attempted to create images that denied Euclidian geometry and, rather, illustrated or embodied Einsteinian relativity. But Kemp argued that it is hard to prove that artists were actually describing scientific theory or that they understood new themes in science. (It might be noted here that Einstein himself denied knowledge of cubism and any relationship between the art movement and general relativity. It is also worth pointing out that the painters, at the same time, claimed no knowledge of general relativity.) Scholars believe, though, that cultural connections existed between the two new ways of looking at dimensions and, therefore, that all thinking was linked together on a deeper level. Kemp agreed. For him, some relationship between what artists painted and the scientific world they lived in must exist.

Kemp tended to emphasize the artist's perspective in relation to science, not the scientist's concern with art. When he did look at science's use of art, his discussion of art and science reinforced Berlin's principle – that before the Enlightenment, the cleavage between the two human endeavours was less pronounced. Kemp did not point out that principle, but his conclusions certainly gave credence to the idea. Because artists were scientists and scientists were artists in earlier eras, Kemp was most successful in demonstrating how art and science could be fused in thinking in his discussions about art pieces created in the pre-Enlightenment period. I was particularly struck by how well science and art could be perceived working together in images chosen from the Renaissance period.

Take Kemp's description of Leonardo Da Vinci's drawings of pregnant uteruses, both human and cow. Da Vinci drew the fetus inside and peeled away the uterine walls to reveal the baby. The placenta was illustrated in a similar way. Da Vinci's passion for understanding the physical phenomena he portrayed in these sketches is palpable. But notes written around the drawings reveal a deeper and non-physical meaning. For Kemp, the notes were the key to the real subject matter of these scientific drawings. Da Vinci used descriptive anatomy to convey the idea that life is governed by certain schema and that all life shares, as a result, the same profound needs. In Da Vinci's words, "the same soul governs these two bodies [cow and human], and the desires and fears and sorrows are common" to all creatures. All mothers are driven by the same love. Kemp elaborated more on Da Vinci's thinking in these drawings elsewhere, and he noted in other articles, for example, the artist's concern with perspective and the problem of projecting the three dimensions of an anatomical reality onto a two dimensional surface. Da Vinci employed these artistic devices in order to illustrate how an under-standing of anatomy helps show how bodies work in action.

Da Vinci used scientific anatomy to say something about mater-nity and its role in eternal reality, and here I recognized certain underlying similarities to my dinosaur paintings. I used anatomy too – bones – to convey a human need to understand certain issues: namely, the meaning of death and its relationship to time. While Da Vinci's simple drawings are masterpieces of art and my paintings are not, the resemblance of the approach (anatomy to express philo-sophic or artistic thoughts) and of the fused themes (science and art) struck me as interesting.

Some comments by Kemp illustrated the union of art and sci-ence, even if that was not his intent. He argued that much modern science lacks art. I found this observation curious, in light of the fact that he also seemed to imply that a great deal of modern art didn't appeal to him. To me the two thoughts were related and could help explain the state of modern culture's approach to reality. His first statement might suggest why he didn't like modern art or what he found missing: it no longer reflected science. Perhaps that is part of the reason it seemed less compelling.(By 2005 I believe there has been some return of science to art.)

Much of Kemp's discussion around individual pieces of art appeared to reveal how artists used science as subject matter. That focus was quite different from the thought I seemed to be con-cerned with – how artists employed scientific and artistic thinking together, in order to create art. When Kemp examined how modern science used visual images, aesthetics was often reduced to illustra-tive motifs. For example, he described mammography in terms of techniques to make images that are "readable" to trained radiolo-gists. While the "readability" problem of medical imagery com-mands a great deal of scholarly attention, I'm not sure I would call these pictures art containing aesthetic themes. Such pictures are using the eye-brain coordination mechanisms set up in our minds to display a scientific phenomenon, not to evoke aesthetic response. Visual images can portray artistic concerns and also illustrate science, but the two themes are not necessarily analogous to each other. Art used to reveal science is not by extension intended to display aesthetics. Science can be employed to elucidate art, but that does not make the art science. Kemp's look at art in the service of science did not explain to me how a unity of artistic and scientific vision worked in creativity.

The dichotomy set up by the idea that paintings using science could be viewed as either works of art or as works of science did not escape Kemp. He pointed out the difficulty in his discussion of meaning behind watercolours portraying deformed beetles (and executed by an artist trained as an illustrator of zoology). Scientists dismissed the paintings as meaningless to science. Perhaps it was the artist's profession as an illustrator of science that made scientists feel compelled to assess the work within the world of science, and not that of art. The labelling itself reveals how deep the cleavage between art and science can be in modern thinking. Kemp did not believe that evaluation of the paintings as either art or science was to the point. "I would resist any attempt to classify it rigidly as 'good science' or 'bad science' – or even as 'good art' or 'bad art,'" he stated. "It is, like all 'good imagery,' designed to enhance our ability to look and think."

Then I realized why Kemp's writings had interested me so much. They allowed me to identify what I believed was the critical issue in the specific art/science problem. It was not the meaning of shared motifs, or even art's influence on science (and visa versa). Knowing where the interface between art and science existed was the difficulty. Scholars are aware that we do not understand what constitutes a demarcation between art and science well, but they believe it is important to do so. As David Topper stated recently, "the very nature of what we mean by 'science' and by 'art'" is at stake when we attempt to understand "the demarcation of science, pseudo-science, and art." For Topper, science does not have a monopoly on empiricism, and art does not have a monopoly on aesthetics. People rarely attempt to find, or locate, the interface, though, between the two areas of thinking. Artists and scientists often collaborate in research on the problem of creativity, central to aesthetics and empiricism and therefore apparent in both art and science, but they do not look for the interface. I think the interface might help explain creativity itself, which could be described as a form of aesthetic and empirical thinking combined. It was, actually, the interface that Kemp described in the art of Da Vinci. Vesalius, Dürer, and other Renaissance artists also worked seamlessly across it.

~~~

The work of a nineteenth-century American artist named Abbott H. Thayer serves as an example of how seamless the interface between art and science can be in more modern times. Through his knowledge of art, Thayer discovered why so many animals have dark fur or skin on their tops and white underbellies. To see how he came to his conclusions, we must understand a certain painting principle: the law of interchange. In paintings, we appreciate the shape of forms through the depiction of value, or the light and dark of lit and cast shadow sides of an object, far more than we do through hue. But the light/dark relationships, or value structures, executed in art pieces are also designed to explain forms found in nature in another way – their shape in relation to atmospheric depth. When light and dark aspects of objects in paintings are made to present what is known as the law of interchange, we perceive not just form but also depth around and behind the forms pictured. In other words, we comprehend an atmospheric space in the painting which is consistent with the way our two eyes, acting together, see the actual world. To make forms in a painting suggests depth and space against or behind that form (regardless of the value of its colour), there must be areas of them that are both darker and lighter than their background. In other words, for a white object to appear to have depth against a black background, that object must be painted darker in places than the background it is seen against. White will have to be presented in certain areas as "black," and black has to be seen as "white" in other places. This variation can be difficult to paint, but it must be done if we want perceptions of three dimensions and atmospheric depth consistent with nature to arise as we view the art.

Scientists had long wondered at the dark and light patterns that so many animals exhibit. The value patterns presented by the animals seemed to act as a camouflage, but how? Scientist assumed that it worked as follows. If the animal was viewed from above, it would match its background, and, if viewed from underneath, it would match the sky. It was Thayer who realized that the camouflage effect was actually much more subtle. By being white underneath and dark above, animals make it hard for the law of interchange to act, with the result that the creatures seem both to lack three dimensions and also to be flat against their "background." When light falls on such animals, their lighter underside would only darken as a result of shadow to a value that equalled their upper side. The blending of values in this fashion would make it less likely that areas of the animals could be both lighter and darker in places than the ground behind the creatures. The result tends to be an image that lacks volume of form and atmospheric depth. These qualities would fool predators, who naturally perceived the world – animals and the land around them – in both three-dimensional terms and depth of space. The dark-on-top and light-on-bottom creature did not look as if it were part of the natural world. The animal did not seem to be alive. Such an image would not, as a result, suggest that prey was in sight.

I can fully appreciate how this camouflage works, because I encountered this flattening effect of form and atmosphere when I attempted to paint pictures of Hereford cattle on summer pastures. The beasts have dark, red-brown backs and sides, along with white tummies. When sunlight falls on the animals from above, their white undersides often seem to become shadowed in a way that makes these areas match the sunlit dark tops, in terms of light/dark. The sunlit tops, of course, lighten in value. It was hard, as a result, to make the cattle look as though they had volume in the paintings, and more particularly to suggest atmosphere and depth against the background of grass and sky. I would try changing the hue of the shadowed tummies from bluish to greenish, but often that, too, didn't work. No matter how I altered the hue, to compensate for the equalizing of value due to the white shadowed undersides and red-brown-lit tops, or how much I described in paint the anatomical form of the animals – their muscle and bone – they often persisted in looking like cut-outs, pasted to the field and trees. Dark-on-top and light-on-bottom animals seem to present in living form a type of demarcation, or interface, between art and science.

The demarcation problem between art and science can be seen most clearly in art emanating out of science. We often find it hard to define what scientific illustrations are – art or science – because of this problem of locating that interface. Interesting art/science issues emerge, for example, when artistic work on dinosaurs is assessed. Many pieces of art and/or illustration have been done over the last century with respect to dinosaurs, and there have also been exhibitions of that art. Within that framework, the work of Charles Knight, in particular, reveals aspects of the way the interface can work.

His paintings illustrate how, years ago, people believed that dinosaurs lived. Between 1894 and 1934, he executed many pieces of art – some large murals included – for the American Museum of Natural History in New York. He also painted for the Smithsonian Institute and the Carnegie Museum of Natural History, the Field Museum of Natural History, and the Natural History Museum of Los Angeles County between the 1920s and the late 1940s. His visions of *Tyrannosaurus rex* and the hadrasaurs (often lumped together as "Trachodon" at that time) bring the beasts to life. What makes the persisting power of their portrayal so interesting is the fact that, today, we believe he had not positioned the animals properly. More modern dinosaur paintings place tails and legs completely differently: we now believe that many so-called bipedal species used all four legs for getting around. It is no longer assumed that heavy tails dragged along the ground. Instead, most scientists argue today that tails were normally carried in the air. Walking on four or two legs is usually immaterial to scientific attitudes concerning tail carriage. Even beasts (such as *T. rex*) that continue to be considered as bipedal held their huge tails off the ground, or so present-day paleontologists think. Knight was, of course, following contemporary scientific thought at the time, and therefore his representations of dinosaurs illustrated what was scientific truth for people then. (And yes, he did depict *Lambeosaurus*, and all hadrasaurs, as bipedal, dragging a huge tail.)

Today, Knight's art can no longer be considered good science. Somehow the "incorrectness," though, does not detract from the compelling power of Knight's images, or from the general respect they can still often command from scientists. The enduring strength of his work results, I think, from the fact that he managed to find an interface. Knight painted seamlessly across demarcation – he eliminated that line, even if he did not clearly identify it. That is one reason why the lack of correct science in the images has not completely devalued them. Location of the interface might also explain why his best paintings to this day convey a deeper message about dinosaurs than art executed in recent years to illustrate what is believed to be the more correct versions. We feel Knight's dinosaurs as living creatures, not as reconstructed illustrations.

Knight's art might teach us about how art and science can work together, but the paintings also carry another message that stands against time – or perhaps we should say illustrates time. Beauty shines through them because they show the historic development of human thinking with respect to dinosaurs. In the process, the art tells us something about how science has interpreted or perceived life over time. The images explain human thinking over time, and they are therefore about history.

I thought of my dinosaur paintings again. I had used the subject of dinosaurs differently from Knight, since I focused on bones, not reconstruction, but I had been unconsciously interested in the demarcation between art and science. When I executed the dinosaur paintings, I had tried to sense the interface by painting seamlessly across the narrow boundary constituting a division between art and science. And, in doing so, I had felt beauty in the presence of art and science. The pleasure that the images aroused in me also resulted from a feel for time passage and memory. I saw again the dinosaurs in *Fantasia* fall to their death, leaving bleached bones in the wind. I remembered my father on chilly, dark Sunday afternoons taking me to the museum. Beauty, symmetry, and balance emerged for me in my depiction of bones because the paintings reflected my memories, on top of artistic, scientific, and historical thinking. Perhaps my art was a "collage" of other art – all predicated on the bones. Then I realized that, long before I produced my dinosaur paintings, I had encountered imagery that attempts to cross the demarcation line. I had actually seen art and science work seamlessly in the movie *Fantasia*. The battle of *Tyrannosaurus rex* and *Stegasaurus*, and the death of *Stegasaurus* alongside *Parasaurolophus*, were not made less compelling stories about life and extinction because of lack of scientific veracity. It is worth noting, too, that *Parasaurolophus* and *T. rex* in *Fantasia* were, like Knight's images, positioned incorrectly, by today's standards, over the pelvis. They also dragged heavy tails.

~~~

Dinosaurs and art suggested to me that Mnemosyne and the muses of art, science, and history together explain how all learning evolves through memory of the past, and, in the process, make us feel beauty or revelation. It was Ernst Mach, a physicist/philosopher teaching in the late nineteenth century in Vienna, who turned my focus more specifically to the linkage of history to art and science. He argued that science and history should be fused. He was most interested in how that fusion advanced science. As I read more about Mach and what he wrote, though, it seemed to me that he was actually interrelating art, science, and history. He reminded me that Clio and Mnemosyne are part of a whole in which the art/science problem (or the art and science muses) fit.

Mach believed that the scientist should try to understand the unity of nature (its symmetry and balance, its overarching character-istics), rather than identify its diversities (particularly evident is the study of taxonomy, for example). While science might contemplate nature's complexity, it should look through research for underlying unity in that diversity. Mach also argued that science should be understood within the perspective of time passage, more particular-ly history. He saw knowledge as a striving for economy of thought, a goal that resulted in shifts of human attitudes to nature over time. The evolution of science, for Mach, followed a Darwinian pattern: knowledge resulted from the survival of the fittest and most econo-mizing theories put forward by humanity in an attempt to explain nature. Mach equated "economy" with "simplicity." "Science acts and acts only in the domain of *uncompleted* experience," he wrote in *The Science of Mechanics*, meaning that science works within the frame-work of time passage or history. And science should seek simplicity because it is beautiful.

Mach's views affected the way a great deal of late nineteenth- and early twentieth-century science was actually practised (Albert Einstein, for example, valued Mach's thinking), indicating that, his-torically, philosophy influenced attitudes to research principles in science. Pauline Mazumdar's book *Species and Specificity: An Interpretation of the History of Immunology*, written in 1995, illustrat-ed more specifically just how deeply philosophy could affect sci-ence's growth. Mazumdar looked at the biological, medical, chemical, and physical developments over time that made an impact on the study of immunology. She clearly showed in the process that differing philosophic approaches to research by individual scientists in all fields affected the development of scientific knowledge. Mazumdar also made it evident that most scientists before the early twentieth century were often philosophers as well as scientists. Mach was by no means unique in that respect.

For me, these approaches to science explained how some of the characteristics in the modern cleavage between art and science emerged. The seeking of unity or the identification of the diversity divide fits with the way the modern split between art and science arose out of reactions to the Enlightenment. The unity approach resembled the German philosophers' definitions of art. Friedrich Wilhelm Schelling, for example, claimed that it was art that looked at entireties, or unity. And because art did so, art was the centre of all knowledge. Science could be genius only when it strove to see the whole in which the pieces it investigated fit – in other words, when science practised art. "Art," Schelling stated, "is the model of science, and wherever art may be, there science must join it". The diversity approach described the scientism that so many of the Romantics reacted against. Science searched for systems, and these systems denied the complexity of nature. Systems or laws, many believed, by definition set up diversity rather than wholeness. The unity/diversity debate in science led, I think, to aspects of art being redefined as aspects of science.

I wondered whether it was possible to argue that nineteenth-century science took specific aspects of art philosophy and called them scientific, in order to give art back to science, and thus to heal the divide that eighteenth- and nineteenth-century developments had brought about between art and science. Robbed of systems of unity when art laid claim to them, science turned to what, for many, was art philosophy – it absorbed aesthetic concepts and then redefined them as scientific. In this way, Mach seemed to be fusing art, science, and history together.

Ironically, or so it seemed to me, Kant's thoughts would be the fountainhead of developing ideas about the meaning of both science and art over the nineteenth century. Kant established a philosophy of science in his *Critique of Pure Reason*, which favoured the unity approach used by Mach, but Kant also explored aesthetics and art's association with holistic thinking. His theories ultimately provided the foundations for art philosophies: for example, Schelling's ideas, distasteful to scientists because he claimed that art, and not science, was the pillar of all knowledge, rested on Kantian foundations. Kant, then, seemed to lay down the grounds both to divide and to unify art and science, and for future scholars to make art science or science art, and perhaps to use philosophy as either one. How do scientists relate to philosophy today? I began to think that a substantial loss of philosophy in science explained a lessened aesthetic approach to scientific research. An interesting line to pursue: Does art remain a part of science only as long as science appreciates any philosophy?

Mach's linkage of economy, or simplicity, and beauty to good science is not today considered essential for rewarding research. But the linkage can still be found in scientific work. Some modern scientists respond to scientific propositions aesthetically, when the idea of simplicity is involved. Martin Gorst's recent study of the development of astronomy, and what might be called its sister science, geology, in *Measuring Eternity: The Search for the Beginning of Time*, reveals that astronomers and scholars of astrophysics working in the late 1990s were particularly drawn to theory and supportive calculations when these demonstrated simplicity. The simplicity struck the scientists as being compelling and, therefore, likely truthful, because simplicity made such theory beautiful. Here again we see the human propensity to see beauty as truth. Beauty, or aesthetics, played a role in convincing these researchers that their work was "scientifically" correct. While today most modern scientists dismiss Mach's other underlying theory – that history is vital to science – not all do. Stephen Jay Gould serves as a notable example of a scientist who believed that the study of history, and, contingently, culture, is intertwined with the evolution of new scientific knowledge.

Dinosaurs allowed me to see that science and art were intimately related. Dinosaurs also suggested that both memory and a sense of history interplayed with art and science, but for some reason I did not grasp the full significance of that thought from my study of the extinct beasts. It would be Mach who really forced me to look more closely at how history could work with art and science. I began to wonder if using art (or aesthetic thinking) and science (or empirical thinking) without incorporating memory and an appreciation of time passage (or historical thinking) resulted in a lack of perception wholeness. Richness results from the appreciation of art, science, and history together. But how does history actually interface with art or science? It was time, I realized, to look more closely at Clio herself. In fact it was more than time. Even though I had by this stage been writing academic history, I had not grasped the fullness of Clio's nature.

All ages are in thy keeping, and all the storied annals of the past [reside in Clio].
    ~ from a Latin epic

Clio … to thee, O Muse, has been vouchsafed the power to know the hearts of the gods and the ways by which things come to be.
    ~ from a Latin epic

CHAPTER FIVE

History

When I met Mach, Clio was not new to me. I knew something of this muse's craft. I had returned by that time to university to pursue the academic study of history, after reaching a mental crossroads with respect to art. I stopped painting because my technique appeared to have advanced beyond my vision. I didn't realize it at the time, but I was not able to portray the ever-larger themes that consumed me because I was not thinking holistically. By looking primarily at the role of aesthetic and scientific approaches to painting, I had not properly absorbed one significant lesson of the dinosaur bones: the equal importance of historical thinking, something different from pure memory, to learning. My inability to see the linkage of art, science, and history when I entered Clio's world also made me turn away from the general problem of art's relationship with science.

Training in history was not new to me when I turned my back on art. At the beginning of our marriage, I had earned a master's degree in history from the university I now attended, the University of Toronto. Although history had been my primary area of education as both an undergraduate and a graduate student at university in the 1960s, I had left the discipline in 1970. There were a number of reasons for this action. To begin with, history's scope was rather limited at that time: it focused on the political past. Although I found political history interesting, it never commanded a real passion for learning in me. There were other reasons why I decided to leave history, though, and they related to my gender. In the late 1960s, graduate school often provided a difficult environment for women, and I, quite simply, found it frightening. I felt unwelcome too, but did not understand why. In some ways I experienced much of what Jill Ker Conway described so clearly in her book *True North* about the same history department, but I lacked both her courage and her ability. Furthermore, my areas of research would require me to live either in London, England, for a year (to pursue the history of the British Empire) or somewhere in the United States (to look more at the history of the American South), if I were to continue my studies at a higher level. Such options were unthinkable for a married woman, and I had no interest in going away either. I wanted to be with my husband. I had fully accepted, in other words, the unspoken dichotomy embedded in the thinking of many middle-class young women in Canada in the mid-twentieth century: that while we were expected to be educated, we were not expected to use our education. Disoriented, I did not feel I belonged in the world of academic historians. I abandoned history, but felt at the same time, vaguely, that I had lost something and also left something undone.

With my sense of art in disarray, I began to think again about completing what I had not finished. And there was time now, in a way there hadn't been earlier. My children were near the end of high school. My father had died. The spring following his death, I put in an application for the doctorial program. Somehow it seemed that I needed to do this for him as well as for myself. And so I became a graduate student once again, after a gap of over twenty years.

When I began the doctoral program in the history department at the University of Toronto, I thought I had abandoned not just painting itself, but aesthetic thinking as well. Conceptualizing as a historian seemed to demand objective thinking of a singular and definable nature: reasoning that related to a scientific, not an artistic (or even a combined scientific and artistic) attitude. Pursuing research for a thesis, teaching undergraduates, and participating in the university environment generally, with its stimulating talks, books, and people, soon made me realize that while history might demand the objectivity associated with a scientific approach, the discipline profited from aesthetic sensitivity. After I completed a PhD in 1997, I became more aware of art in history. My first book, *Ontario's Cattle Kingdom: Purebred Breeders and Their World, 1870–1920*, published in 2001, showed me that an objective methodology was enriched when I looked at historical problems from an aesthetic point of view as well. Pauline Mazumdar, the historian who wrote on immunology and a professor with the Institute for the History and Philosophy of Science and Technology at the University of Toronto, helped me see this connection. After reading *Ontario's Cattle Kingdom*, she described purebred breeding as "the art of genetics."

One beautiful July morning I awoke and knew I had another book to write. I would deliberately attempt to identify aesthetic and scientific forms of thinking in historical reasoning this time, to understand better what historical conceptualization meant. The book would focus on the way characteristics of purebred animal breeding have developed since 1800, and would make use of all the information I had collected over a lifetime on Shorthorn cattle, Collie dogs, and Arabian horses. I would assess the culture of breeding strategies, the economics of purebred breeding, and the ideology of breeders. I would include images from art, and discuss the role of art in breeding. I decided to weave literature I had read in childhood – the work of Albert Payson Terhune and Walter Farley in particular – into my story. I was lucky. The breeds that so interested me provided critical information, which allowed for a coherent story about historical breeding. Through the writing of this history, though, my memory and nostalgia made me want to understand the past of Shorthorns, because I loved the animals on the farm; the past of Collies, because of my memories associated with Terhune's beloved dog, Sunnybank Lad; and the past of Arabian horses, because of childhood dreams based on Farley's descriptions of a fictional black stallion. I hoped, essentially, to distill my aesthetic nostalgia into something meaningful to others through the writing of history, not the painting of paintings.

The book, *Bred for Perfection: Shorthorn Cattle, Collies and Arabian Horses since 1800*, was published in 2003. Working on it taught me that the writing of history could be a vehicle to express, above and beyond the apparent subject matter at hand (in this case attitudes to animal breeding over two hundred years), something about how people think. A description of historical animal breeding could reveal a deeper story – the interaction of memory and the passage of time as expressed in history, as well as art and science in knowledge. The pursuit of academic history had introduced me to the importance of Clio herself in holistic thinking. In the end, Clio, or history, would show me that, through historical thinking, art and science work together most productively, and that Mnemosyne aids all three facets of knowledge.

Now, at last, I had grasped the lessons of the dinosaur bones and Mach's teaching: we cannot really appreciate meaningfulness in either art or science without addressing history. Nor can we understand the implications of history without combining artistic and scientific approaches to any problem relating to history. Art, science, and history all need art, science, and history together. Mnemosyne needed Clio to disperse wisdom through art (the need to relate the self to the outside world, for example, through an understanding of symmetry and balance and the desire to find beauty, however it is defined) and through science (the drive to think objectively about nature around us in order to comprehend physical laws outside our control). Clio heightens knowledge in both art and science. And art and science enrich Clio.

I had thought a great deal about the meaning of both art and science. Now I began to contemplate the meaning of history – the nature of Clio, not her work. History is, of course, the objective attempt to learn about past societies of humanity. The discipline had expanded its scope hugely from the time I had been at graduate school in the late 1960s: it now explored in considerable depth the evolution of all human thought with respect to any activity. Time passage remains at the heart of any such study, though, and therefore of history. I thought: somewhere within the mystery of history lie answers to the larger questions of the relationship of time to knowledge, and the importance of perceiving art and science as a unity. Perhaps the history of how history developed might help me unravel with better clarity these mysterious connections between art, science, and history in reasoning. History seemed to play a critical role in the ability to think holistically. I began to read more historiography in order to see how the methodology for studying history had evolved. I wanted also, on a less-complicated level, simply to know more about the background and development of my new profession. I wanted to understand history in somewhat the same way that I had earlier tried to understand art.

~~~

What role, I asked myself, have artistic and scientific approaches played in a developing methodology for the study of history? How have art and science affected approaches to historical thinking? Perhaps if I understood the way artistic and scientific reasoning had been interwoven with historical thinking over time, I would better see why, today, it is so easy to think that history allies itself closely only with science; why so many scholars see this humanity in "scientific," not "aesthetic," terms; and why some modern historians even believe that the study of history equates to that of physics, because both use mathematics. Yet the writing of my two academic history books had indicated otherwise to me.

After a great deal of reading, I developed a synthesis that, for me, explained the development of Western history and allowed me to relate the discipline to some of my other lifelong concerns: the interaction of art, science, and memory in all thinking. Because this synthesis in the end would lead to a final or more encompassing way of understanding how holistic thinking works, a description of how I arrived at the synthesis is important to this story of a pathway.

The history of history showed me that the beginnings of modern Western history's approaches to the subject lay in Greek thinking. While Mnemosyne gave birth to history, Greco-Roman thinkers together laid down the basis of a subjective (in this case artistic) and objective (here scientific) approach to the study of history. Even though they often tended to position the discipline in the genre of literary art, designed to entertain and provide moral enlightenment, the importance of evidence to historical writing was emphasized by Herodotus, clarity of thinking by Thucydides, and the need to understand causation by Polybius. All the above men, and Cicero as well, wanted history to present an unbiased view in its assessments of events. For the ancients, whether a methodology was based in art or in science, history must be founded on what was believed to be the truth. Cicero called history "the light of truth." History, then, was irrevocably interconnected to both artistic and scientific forms of thinking from its academic beginnings in Western civilization.

The discipline's methodology began to take on its modern shape as a result of intellectual ideas that arose over the eighteenth and nineteenth centuries. The Enlightenment drove history and its concern with the past into the background and denuded the discipline of both scientific and aesthetic reasoning. For the philosophes, history fit into what might be described as popular culture. The Enlightenment's fascination with science tended to make people dismiss history as fantasy or mythical tales that had nothing to do with real knowledge. Because historical methods could not clearly be identified as scientific, most French thinkers, including Descartes, stated that history lacked any real claim to veracity. Voltaire believed that history could teach us about the heights of human achievement, when it dwelt on the rare periods of human advancement, and, under these conditions only, was history worthy of attention. It then had the power to teach. All other attempts at history were exercises in amusement.

But within this period, history still managed to provoke passionate defenders who valued its study as an intellectual exercise of considerable worth. One of the earliest was Giambattista Vico, who believed that the only real knowledge we can have must be generated from an understanding of humanly constructed subjects such as history. Vico argued that history revealed what men and women have made of the world in which they find themselves. He held that the process of history explained humanity's interaction with nature over time. There could be a study of the mind, which was the history of its development: history showed how ideas evolved, and that knowledge was not a static thing made up of eternal, universal, clear truths that result from laws.

The reactive Romantic period brought the discipline into the limelight again and stimulated a fundamental shift in approaches to history. It was the special attitudes of the Romantic artists to the past generally which brought that trend about. These people associated the past with subjective introspection, but their obsession in itself with the past stimulated an interest in it for its own sake. In 1994 Raphael Samuel, in *Theatres of Memory*, described the subjective introspection in Romantic approaches this way: "Romanticism built on time's ruins. Its idea of memory was premised on a sense of loss." It is worth quoting Wordsworth's poetry to illustrate how strong the sense of time passage as decay, past being lost, and introspection through contemplation of a personal past can be in Romantic thinking. The poetry also reveals that concern with the idea of "past" coloured these visions as well. In "Intimations of Immortality" he writes:

There was a time when meadow, grove, and stream
The earth, and every common sight,
To me did seem
Apparelled in celestial light,

The glory and the freshness of a dream.
It is not now as it hath been of yore;
…
But trailing clouds of glory do we come
From God, who is our home:
…
The thought of our past years in me doth breed
Perpetual benediction …

In "Lines: Composed a Few Miles above Tintern Abbey," Wordsworth has this to say on the subject of the past:

"These beauteous forms,
    Through a long absence, have not been to me
As is a landscape to a blind man's eye:
…
The picture of the mind revives again:
While here I stand, not only with the sense
Of present pleasure, but with pleasing thoughts
That in this moment there is life and food
For future years.

The poems say that it is memory of past beauty and nature that saddens or fills the heart with nostalgia because of the loss of beauty and, with it, purity. Only through appreciation of the past do the present and the future make sense. We can understand nothing about beauty today if we do not approach the meaning of the past. While the poems show an introspective passion for the past, a hidden desire to understand it on its own terms can be felt in the words too. The Romantic concern with bygone eras allowed for a study of those periods detached from a nostalgic search. Subsequent interest in the values that the past offers to society triggered a new appreciation for the discipline that examines the past – history.

Under this stimulation, the discipline separated itself from its eighteenth-century popular culture roots by initiating a diachronical (viewed within the time frame of the happenings) as opposed to anachronical (viewed within light of the present or other ages) approach to the analysis of bygone eras. Earlier times could, as a result, now be viewed without depreciation. For the first time in hundreds of years, the Middle Ages would be valued in their own right. The transition to a diachronical attitude in history was more or less complete by the late nineteenth century. The attachment of anachronical thinking and myth preservation to history had largely passed away. Leopold von Ranke, a German historian, led this trend, and he is accredited with being the father of today's academic history.

While shifting attitudes to history over the eighteenth and nineteenth centuries might have brought about the move to a diachronical stance in the discipline, my reading convinced me that it was the way the methodology for studying history changed over the twentieth century, and more rapidly in the latter half of that century, which resulted in the growing emphasis on scientific approaches to historical research so evident today. The evolution of the social sciences in the second half of the twentieth century would increase scientism in all the humanities, but especially in history. Sociology, in particular, had a profound effect on history.

I found the early twentieth-century words of Bennedetto Croce particularly prophetic of what was to come in greater force later that century. He stated that sociology "would work at a *mechanic* of history, a social physics," and enlarged as follows: "A special science arose, opposed to the philosophy of history, in which that naturalistic and positivistic tendency became exalted in its own eyes: sociology. Sociology classified facts of human origins and determined the laws of mutual dependence that regulated them, furnishing narratives of historians with the principles of explanation, by means of those laws. Historians, on the other hand, diligently collected facts and offered them to society, that it might press the juice out of them – that is to say, that might classify and deduce the laws that govern them." Croce concluded: "Hence a vicious circle, evident in the connection between history and sociology, each one of which is based upon and at the same time to afford a base for each other. But if, with a view of breaking the circle, history be taken as the base and sociology as its fulfillment, then the latter will no longer be the explanation of the former, which will find its explanation elsewhere." I would argue that to "press the juice" out of facts is to deny the aesthetic aspects of those facts – their impact on other facts.

By the 1970s, sociology would increasingly encourage history to rely on a scientific approach in its study of the past. Quantification, a tool of sociology, came to play a greater role in historical research with the expansion of interest in social history. Quantification is the use of data to explain historical events and, at the heart of it, of course, is mathematics. I have said earlier that mathematics is a part of both art and science. When this type of mathematics is applied to history, though, it does not seem to relate to artistic approaches, or emanate out of art. When a quantification emphasis alone underpins the study of history, it is only numbers, resulting from the compilation of data, which are used to elucidate events in the past or what people did or even thought historically. Art never looks at human activity outside the framework of human thinking, even if art uses mathematics to organize its shape. Quantification methods in the study of history reflect science's use of mathematics; the increased use of quantification in the discipline, therefore, tends to push history further away from aesthetics. The heavy emphasis on quantification in historical research meant that many of the books I read in my years as a doctoral candidate were based on this mathematical approach. No wonder I had found the artistic aspects of history to be so hidden or masked when I began an academic career in history.

The quantification emphasis triggered something of a backlash in the thinking of many historians by the late twentieth century. Some began to stress the importance of the other traditional method of historical research – qualification. Qualification in historical writing results from the historian's assessment of material that provides evidence of human thinking, usually found in written documents. Visual images and the spoken word are also used in the qualification approach. Sources of this nature reflect attitudes to reality, not reality itself. The historian has to understand that qualitative documentation results from humanly constructed ideas. All such documents also tend to reveal the past's biases, and the present's as well, even to supposedly objective "facts." In short, such sources display myriad meanings that show their association with beliefs emanating from both the past and the present. The approach to such documentation, then, requires careful thinking that should be diachronic in nature. What is past ideology has to be separated from what is present ideology.

My exploration of how history developed between the eighteenth and end of the twentieth centuries uncovers why history today is closely related to a scientific outlook, and even science itself. The quantification method of research clearly allies history with science. Since Mach and the dinosaurs had shown me that science was associated with history, and because my historiographic research had revealed a link within history to science, I had now connected both science to history and history to science. Earlier I had interconnected science and art with each other. In order to see underlying unity in thinking – scientific, artistic, and historical thinking working together – I needed to find an intimate relationship between art and history. Did modern developments in history "press the juice" of art out of the discipline? Had my look at the history of history revealed any presence of art in modern historical research? Here I ran into problems.

The quantification/qualification division of methodology, which might clearly reveal science in history, did not do the same for art. Quantification and qualification even had a strange way of crisscrossing each other from the point of view of artistic or scientific outlooks, through the process of diachronic/anachronic thinking. We should not see, as I had at first, a diachronic approach as fundamentally scientific, or an anachronic approach as artistic. Nor should we be tempted to link diachronic thinking with quantification, and anachronic thinking with qualification, which for some reason is also easy to do. Understanding how quantification fits with qualification, and how a diachronic replaced an anachronic approach in history apparently did not explain aesthetic thinking in history. I could not, therefore, simply equate quantification with science and qualification with art. The history of research methods applied to modern academic history, then, did not reveal a presence of aesthetics in the discipline.

Perhaps I can clarify the fact that quantitative/qualitative, diachronic/anachronic issues cannot be said to reflect the root of aesthetics in history by showing how they played a role in my own research. The focus of my doctoral program was Western agriculture, with a particular emphasis on livestock production in Canada over the late nineteenth and early twentieth centuries. There has been a fairly long tradition of looking at agricultural production historically through a compilation of numerical data. Census numbers have been used, for example, to reveal how many acres in Canada West (later Ontario) at a particular time were planted in wheat versus fodder crops. Data tells us too what wheat yields per acre totalled over the years, and what percentage of farm income came from wheat over time, versus beef, pork, or dairy products. It is clear from the census figures that farming in Ontario moved away from wheat and to livestock farming over the late nineteenth century, but many questions around that move were not explained by those numbers. I found myself looking for answers to some of those questions. I wanted to know, for example, not just why farmers moved to livestock (and pastoral) farming over wheat (or arable) farming but also what their strategy was in relation to the products that livestock was capable of yielding. What, in other words, were farmers thinking when they attempted to mould a biological product? What did they expect to achieve when they worked with cattle?

By 1870 beef had become a mainstay of the agricultural economy in Ontario. But as the century went on, there would be a shift towards dairying, particularly the manufacture of cheese. The cheese industry has received a great deal of scholarly attention, but virtually all of it has focused on the economy and technology that relates to cheese manufacture (or on the shifting gender control from women to men in it). Cows, critical to the making of cheese, have commanded little attention from historians. One reason, perhaps, is that though the census provides data on cows, there is little information within those figures about the ability of the animals to provide agricultural products. Cows are not machines that produce equally. This variability seemed to illustrate the crux of the problem to me. Since cows explained the functioning of the strength of the cheese industry as much as did factories or numbers of pounds made and exported, and figures could not adequately explain the relative ability of the cows to support dairying, we must look outside census data for answers. I asked myself two questions in relation to these facts: How did beef cattle farming relate to dairy cattle farming, since cattle were the basis of both? And how did farmers make cattle serve each industry? I found – by reading what agricultural experts said, by learning what actual animals on farms looked like from qualitative information arising in the farm press, and by then relating this information to available census data – that when farmers increasingly emphasized dairying, they effectively made beef farming part of dairying. Beef came to be seen not so much as a commodity in its own right, as it had been earlier in the century, but rather as a byproduct of dairying. The growing specialization of the dairy industry for cheese was not evenly matched with a developing specialization in the cows to produce that commodity. By 1900 the dairy industry was, as a result, more closely attached to the beef industry than it had been in 1870.

I had used both quantitative and qualitative research to form my conclusions and had done so within a diachronic approach to each. Empiricalism breathes through my analysis of numbers and my relationship of numbers to the testimonial documents, but where was art? Had history's ancient aesthetic roots died? Had it vanished over the years? No. Art had not vanished from history, I knew from my evolving pathway. I had experienced an ongoing connection between art and history in my own painting and writing. I had found history within my dinosaur art work, and my writing of history had convinced me that art still breathed through the discipline. Somehow history's aesthetic roots had become masked by the more recent developments in academic history. Perhaps if I understood how art and history actually interfaced with each other, I would better locate aesthetics in history. In spite of my historiographical research, my sense that culture and science were irrevocably entangled with each other over historical time, or my appreciation of the way the quantitative/qualitative and anachronic/diachronic diamond worked in the discipline, I had not pinned down the aesthetics roots of history. Could aesthetics be found in the way I fused quantitative and qualitative material in order to find the answers, which might be said to represent balance?

~~~

A balance between quantification and qualification: I realized, then, that the very existence of the quantification/qualification debate in historiography was an important key to the solution of my problem. The overemphasis on a methodology, which implies that scientific laws govern human activity, might have made historians look more closely at a non-data-based approach to historical thinking (or an approach that did not rely entirely on a scientific outlook), but in the process something else emerged. The fundamental belief that history was not really a type of pseudo-science took hold in many academic circles. Historical thinking, scholars recognized, was composed of more than science. The desire to return to qualification also seems to suggest that a critical balance, or special tension, between different ways of thinking is inherently present in history. The discipline appears to strive instinctively for a special blending between two approaches. That special blending might be described as the Goldilocks syndrome: not too little and not too much of either form of reasoning, but, instead, just the right amount of both. An inclination to argue that any methodology used for research in the discipline should reflect only science had disrupted a sense of that blended syndrome operating in the study of history. How, then, could one describe the other type of historical thinking within that critical tension? Was it art? In order to answer these questions, I took a slightly different line of reasoning.

I asked myself again, What were the underlying reasons why I seemed to value history? The answer related to history's concern with the effects of time on thinking and human activity. I recalled my sense of wholeness in my writing of history. Concern with time and wholeness in perception. Balance. The need to understand time better, I now saw, had even played a role in making me turn to the study of history. I had not learned enough about time from my efforts to appreciate the power of memory, the meaning of art in relation to science, or the nature of extinction. The seat of art in history might lie here – in the discipline's basic approach to the meaning of time itself. I wondered: Could I link time in history to time passage as collage? Quite quickly I learned that I seemed to be justified in doing so, because others had made the connection.

Historians, Ernst Breiach argued in his *Historiography: Ancient, Medieval, & Modern*, written in the 1980s, manage to show people that "human life is never simply lived in the present alone but rather in three worlds: one that is, one that was, and one that will be." Breiach explained historical thinking as an appreciation of time as a collage. He said: "In theory we know these three worlds as separate concepts but we experience them as inextricably linked and as influencing each other in many ways. Every important new discovery about the past changes how we think about the present and what we expect of the future; on the other hand every change in the conditions of the present and in the expectations of the future revises our perceptions of the past. In this complex context history is born ostensibly as reflection on the past; a reflection which is never isolated from the present and the future. History deals with human life as it 'flows' through time."

Perhaps even more convincing are the words of Jose Ortega y

Gasset. History is the study "of the present in the most vigorous and actual sense of the word." He continued: "The past is in truth the live, active force that sustains our today. There is no *actio in distans*. The past in not yonder, at the date when it happened, but here, in me. The past is I – by which I mean my life." History, for Ortega, was the dispassionate study of humanity, which could be explained through an assessment of time's impact on his own mind and heart. To comprehend anything relating to our humanity, Ortega argued, we must look at that group or person's history. "Man lives in view of the past," he stated. "Man, in a word, has no nature; what he has is history." Ultimately, "life only takes on a measure of transparence in light of *historical reason*." For Ortega, only when time is collaged through a historical outlook do we understand what we are.

Breiach pointed out that history dealt with human life as it flows through time, but he saw time itself as an endlessly varying form of collage. Ortega claimed that history was the study of both the present and the future because of its concern with the past. Flow of time, but time also as a collage. These were the clues I needed. I could link them to what my pathway had taught me. I thought: history actually is, simply and purely, an intellectual study of time passage as a collage. And time viewed as a collage, I had learned through various avenues, implies the presence of art. I had discovered, for example, that aesthetic nostalgia, a fusion of past to present via memory, fuelled art. I had an answer now for where art lay in history: it is the way history studies time that brings aesthetics to it. History's heartbeat, therefore, not any aspect of research methodology, is the main fountainhead of the discipline's art. When I saw history's heartbeat as art-driven, I could even understand better how science resided in history or breathed through some of its research principles. For example, the idea that time in history is aesthetic is not necessarily in conflict with the sense that quantification brings science to history. Data can be used to *reveal* human activity, even if that scientific, mathematical material cannot *explain* that activity by establishing laws that would do so. I was beginning to sense the way wholeness worked through history.

Time can be described as art, but when can it be described as science? The question was important to answer if I was to be sure that the meaning of time in history was art. How is the impact of time assessed from a scientific point of view? I asked myself. And what, in turn, would it mean to say that history studies time with scientific reasoning? Science seeks to understand natural laws that govern the dynamics of processes. If history searches for laws of this nature – laws outside human control – in order to explain human behaviour through the passage of time, we would be encountering a purely scientific approach to time in history. If history shows predictability through such laws, we would find science. Do we expect to find predictability in the way historical events have unfolded? Do we look at time in history (with any methodology) in a fundamentally similar fashion to the way we appreciate time in relation to outer space – for example, light years, which describe distances between galaxies or stars? Is time in history the same as time in the theory of general relativity? Are we thinking historically when we look for laws that explain the way events take place over time? Do we believe that mathematical data can tell us why historical events have happened the way they did? I read further and tested what other scholars, in history and outside history, had to say about history's study of time in relation to these questions.

Everything I looked at suggested that philosophers (and not just philosophers of history) and historians supported my contention that art, not science, lay at the bottom of history's assessment of time. Scholars tend to describe the way history studies time in one of two ways: either as a study that shows time passage to be aesthetically collaged through the teaching of personal memory or as an attempt to understand time's interaction with changing human thought – time's interaction with random events that, in turn, shape subsequent happenings. Historical patterns show not only unpredictability as far as these scholars are concerned. Historical happenings also cannot be separated from human thinking and activity. None of these attitudes to time in history suggest the presence of scientific laws. Sometimes scholars use the word "science" in relation to their discussion of time and history. But when they do, it seems that they are attempting to maintain an aesthetic/scientific or even subjective/objective balance in their overall approach to history. Even if history approaches time passage aesthetically, for philosophers and historians generally the discipline is still structured around aspects of art and science in accordance with the Goldilocks syndrome.

Jacques Le Goff wrote in *History and Memory*, "Memory is the raw material of history" and "History transforms memory into knowledge." Le Goff's point is that history remains objective, even though it looks subjectively and introspectively at human feelings about the meaning of time. Samuel put it this way. "History has always been a hybrid form of knowledge, synchronizing past and present, memory and myth, the written record and the spoken word." Time, for history, presents a collage of past, present, and future, one that the human mind understands aesthetically through an appreciation of the power of memory.

Many go out of their way to separate history's view of time from that of science, in order to show how historical reasoning is able to supply a unique type of knowledge. Scholars in the fields of mathematics and science not only see aesthetic thought in history and but also appreciate the discipline's explanation of events within the framework of their uniqueness, and through that uniqueness the ability to change the course of future events. Viewing time passage as having random effects on events, they believe, offers overarching vision to important problems, and in a way that science, which looks for continuity and predictability in laws, can not. Philosophers of mathematics often give history an exalted position because history tries to understand time within the needs of the human spirit. History's sense of time, for mathematician C.J. Keyser, was critical to the evolution both of human culture and of all acquired knowledge. H.R. Rickman, a philosopher writing in the year 2000, also saw art in history and valued the discipline. He claimed that history and historical judgment alone could explain the nature of humanity. History supplies a meaning for time that the sciences and social sciences do not, and only from history, Rickman argued, will we learn about the depth other avenues of knowledge can provide and the way all knowledge interfaces with us.

The scientist Stephen Jay Gould believed that the art and humanism in historical thinking were invaluable for science:

> The organizing principles of history are *directionality* and *contingency*. Directionality is the quest to explain (not merely to document) the primary character of any true history as a complex but causally connected series of unique events, giving an arrow to time by their unrepeatability and sensible sequence. Contingency is the recognition that such sequences do not unfold as predictable arrays under timeless laws of nature but that each step is dependent (contingent upon) those that came before, and that explanation therefore requires a detailed knowledge of antecedent particulars. Each complex state has a multitude of possible outcomes, and any alteration early in the sequence sends history cascading into a different, but equally sensible, channel. History is therefore unpredictable before [the] unraveling of a series of events, but just as explainable as any kind of science thereafter.

Often, Gould added, "a denigrated history [is posed as opposite] a triumphant science: mere story vs. explanatory vigor; biased tale-telling vs. objective description; art vs. science." This scenario is detrimental to both history and science, and, through them, art. Gould argued that the sciences of Earth's past – such disciplines as geology and paleontology – would benefit from historical thinking, with its ability to see directionality and contingency in an explanation of what happens over time. Paleontology and dinosaur bones: I thought again of my paintings. For me, they had presented beauty or revelation because I had thought about them artistically, scientifically, and historically. I had used memory to weld that process together.

I had seen "history" in art, and both within a scientific field.

A look at the history of history and at the evolution of its methodology had shown me science in history. Now I had the roots of art in history too – an approach to the meaning of time. Consider my own work, which I described in relation to the beef and dairy industries in Ontario, and Gould's ideas of "directionality" and "contingency" in relation to the study of time's effects. Directionality: my desire to understand the complex thinking of farmers (from the point of view of their breeding practices when they decided to use cattle primarily for dairy rather than beef products) might be described as a study of "directionality" in history. I was clearly more interested, for example, in the attitudes of people to circumstances than in outlining the rise of such circumstances. In other words, I followed "the quest to explain (not merely to document)." Contingency: I found farmers did not behave in what might at first glance be considered a predictable manner. They did not simply start to breed "dairy" cows instead of "beef" cows. They chose to strategize in such a way as to capitalize on advantages obtainable in both industries. Farmers reacted "contingently" to the existing situation because complex and interdependent factors affected them – not the least of which were the attitudes of agricultural experts to animal improvement through purebred breeding and, strangely enough, the rising per capita consumption of beef in both North America and Britain. The specific attempts made by farmers in that era to serve the beef industry through the dairy industry worked in the timeframe of the late nineteenth and early twentieth centuries. It would not work in the same manner today. There was, therefore, "unrepeatability and sensible sequence" in farmer actions at that particular time and in that particular place.

History as "a hybrid form of knowledge, synchronizing past and present," history as a process of memory, as a transformer of "memory into knowledge": it is hard to escape the sense that my interest in the historical cattle industry, and my wish to understand farmer thinking with respect to animal breeding, in some way emerged from the practical aspects of my life which had involved living cattle. Memory and the enrichment of time's meaning when it is viewed as a collage: our cattle breeding activities on our farm in themselves emerged from a collage of memories that allowed me to study larger implications of historical animal breeding, as found in dogs and horses. I was then drawn into related historical issues such as biometry, mendelism, Darwinism, eugenics, Social Darwinism, population genetics, and evolutionary biology. I would see next, from the way other historians looked at meaning behind my work, that in the process of studying these topics within the framework of animal breeding, I was also dealing with certain large contemporary issues. I was, in effect, describing the early patenting of biology, for example, and initial attempts to protect the intellectual property people believed they had created in living, not man-made, things.

I had identified from the synthesis what I needed to sense wholeness from history: the clear presence, not just of science, but of art too. Now I could appreciate that a blending of thinking goes into and creates history, and that artistic, scientific, and historical outlooks can be entangled with one another. Aesthetics helps to explain historical events, for example, but, simultaneously, history might reveal artistic development. At the same time that scientific approaches underlie historical reasoning, science in itself can be seen as a historical phenomenon. Within the mystery of history lies answers to the larger questions of the relationship of time to knowledge, and the importance of perceiving art and science as a unity. A historical appreciation of time, as I had suspected, could actually act as a link, or even as an interface, between art and science.

The following ideas occurred to me. History demonstrates the unity of artistic and scientific thinking in its search for knowledge of the past, and therefore it can show how art and science are linked. History studies time, but an understanding of time develops from combined artistic and scientific thinking. Memory, nostalgia, beauty, truth, love, lack of concealment, mathematics, and subjective and objective thinking are all part of any study of time's impact on humanity, and therefore go into any study of history. Historical thinking, then, is a fusion of subjective and objective thinking that relates to both aesthetic and scientific forms of reasoning in order to learn how the interaction of time with events in turn results in other events. The lack of art in science and the lack of science in art is the lack of a historical approach. That is so because history is the fountainhead behind both artistic and scientific attitudes to the problem of human nature. I would even go so far as to say that the meshing of artistic and scientific thinking is historical thinking. Could I say as well that history results from the attempt to unify art and science, because its vehicle is the study of time? It seems that all holistic thinking results from the union of aesthetic, scientific, and historical approaches: while within art and science lies history, within history lies art and science.

My synthesis of history's development and meaning took me farther down my pathway to greater understanding. It also revealed to me something about myself. It finally explained to me how much and why I had always loved history, loved it even in the difficult years of my master's degree. History was fundamental to all my thinking. I realized now that I had lost some essential part of myself when I had abandoned the discipline in 1970. It had taken all these years to find that part of me again. History, in many ways, lay at the bottom of the way I, as a person, was naturally wired. Would it have been different, I wondered, if I had gone on and completed the PhD right after the master's? Would it have been different if I had never studied history, but instead pursued an education in either fine art or in painting? After all, I have been interested in art from a time before I could print. But if I had followed either of those avenues, would I have sought out information in areas like philosophy, for example, or even historiography, to the degree I did? Is my appreciation of history only richer because of fifteen years of serious painting? And, in turn, was it my approach to art that taught me where to look for the real meaning of history?

~~~

Efforts at understanding the meaning of history, and the role of art and science in the discipline, proved to be critical to the way I followed my pathway. History let me see unity within complexity, and allowed me to make sense of both in relation to each other. History explained to me, as a result, how wholeness works. All subjects offer us the richest meaning when they are approached aesthetically, scientifically, and historically. Any avenue by which to feel wholeness is also useful in understanding more about the aesthetic, scientific, and historical parts. All avenues that allow us to think wholeness are worth exploring. Together they enrich our sense of the way the parts interact with each other to make the whole.

My mother's message, through death, had been the right one for me. My memories were, in effect, about more than just my past. They were about my present and my ability to think. Tools necessary for academic thinking emerged from their crucible. They paved an avenue that could lead to knowledge. My past and my history could also tell me something about the human spirit. Memories encoded my mind with patterns of thinking and shaped what interests would fill my adult life, and also how they would become intertwined with each other. I would not have learned without an appreciation of the power in memories, because it was they that created my passion for learning.

Then again, Zeus loved Mnemosyne, of the splendid tresses, from whom were born to him the Muses, with veils of gold, the Nine whose pleasure is all delightfulness, and the sweetness of singing.

~ Hesiod, "Theogony"

The Muses provided music at the wedding of Cupid (Love) and Psyche (the Soul), who would, as a result, produce a daughter, named Pleasure.

~ Roman myth, Apuleius.

CHAPTER SIX
Shells

S hells: I love collecting them. And I found I could dream science and think art, or I could think science and dream art, when contemplating the entangled aesthetic and biological reality they present. But they seemed to be associated with a sense of time passage as a collage when I thought of them from any scientific or artistic point of view. Shells hold the past because they are the remains of dead molluscan life. I thought of the animals that once lived in them. They are the present when I encountered them in the sand, treasures still wet from the sea. When I search for shells, which I do on every tropical beach I walk on, I listen to the sea and watch the sky's clouds in rhythm with the sea's waves, in order to take those memories home along with the shells. They become the future through this avenue, and also because I will study them. I feel science, art, and time passage fused together when I appreciate shells.

Shells bring science to me because I am interested in them from a biological point of view and want to learn about both the evolution of molluscan life and the science of conchology. The variation in the shells that mollusks build is almost unbelievable. Some of these differences are related to hemisphere location. Mollusks develop different shells in cold water than in warm. Northern species have thin shells with little colour variation. Southern species can be huge, compared to those from northern waters, and grow shells that are often thick and brilliantly coloured.

Because I recognize boundless beauty in them, shells hold aesthetics. They are all beautiful to me; and even shells worn from rolling in the sea, devoid of colour, show a symmetrical beauty that is breathtaking. I am anything but alone in finding that aesthetics breathes through shells. They have stimulated the production of art for hundreds of years. Think of Botticelli's scallop shell in *The Birth of Venus*, painted in 1480! The aesthetics in shells, then, is linked to history – the history of art. Shells have been part of literature too. They were central, for example, to Anne Morrow Lindbergh's book *Gift from the Sea* and to Bachelard's aesthetic assessment of physical space in his book *The Poetics of Space*.

I find stories of human history in shells as well. Cowries played a role, for example, in the economy of various societies because they were often traded as currency. Other shells could be precious objects, like gemstones. It sometimes took a fortune to own one. A Matchless Cone, for example, in 1796 sold at auction in The Hague for five times the amount that a Vermeer painting, *Woman in Blue Reading a Letter*, did at the same sale. Rare shells could be so valued that manufactured fakes would find a market. In some cases today the counterfeits are worth more than the real shell. In the early 1700s, for example, the then rare Precious Wentletrap was imported in small numbers into Europe from China, and it has been suggested that reproductions made of rice paste were as well. If one of these fakes turned up today, it would be priceless because of rarity, while the real Precious Wentletrap (now found plentifully) is worth only a few dollars. Scientific interest in shells over time has proved to be critical not just to the development of conchology but to knowledge in other major scientific fields: geology and paleontology, for example. In the 1820s the great geologist Charles Lyell came to some of his conclusions about the age of the Earth and the dynamics of geological change through his knowledge of shells. Over the nineteenth century many scientists puzzled over the presence of shells high in the mountains, which suggested that ancient lakes had existed there. The study of fossilized shells contributed to the development of paleontology, because it clearly revealed the process of extinction and added to knowledge of evolutionary biology. Shells, then, can explain not just certain historical social patterns but also many aspects of historical scientific thinking as well. Shells can link history and science.

Suddenly I realized that I had discovered a vehicle by which I could try to feel the whole – the presence of science, art, and history in a complex and intertwined way. Shells might present pure science, pure art, and pure history to me, but they also showed how complicated the linkages of all three to each other can be. I found art, for example, when studying science. And here mathematics played a central role. In my biology research I located a demarcation line between science and art, or an interface between the two avenues of understanding, when I realized that the dynamics of biology (science) and the dynamics of art could be found in the same mathematical expression. A crucial tie between art and science through mathematics can be seen in how shells develop. The way the whorls grow around the axis of a shell's body during the life of a mollusk presents a mathematical symmetry that is also fundamental to a form of aesthetic symmetry. That is so because the spiral represents, over and over again, the golden section that has been used by artists for centuries to design paintings. It struck me as interesting that I was experiencing something of the reverse thinking of Abbott H. Thayer, who, by explaining the light-dark colour of animals through the law of interchange, found science through art. I found art through science, when the mathematics in shells showed me the way another interface worked between art and science. I had also, I thought, discovered convincing evidence that mathematics could be used to demonstrate a union between art and science.

Because I could understand shells in this multifaceted way, I was able to see that other undertakings in my life could be approached in the same overarching fashion. Shells, for example, gave me back my desire to create art. I started to photograph the shells I had collected from beaches. Then I found myself painting pictures composed of them. I no longer asked myself whether the work I produced was "good" or "bad" art. It was, quite simply, a representation of my wish to understand. I was now using art as an avenue to learn more about both science and the historical passage of time. Soon I wanted to paint not just pictures of shells but other images as well. The act of painting had heart again for me. I looked, from this new centre of wholeness within my thinking, at all the natural beauty around me. Light and shadow, for example, poured their endless variations of precious beauty over me through my eyes. I appreciated nature now from a historical point of view and also within the framework of my own life.

I approached the writing of history with greater balance, and in the process I came to understand more fully how complicated the interweaving of human thinking could be, made so by the innumerable ways art, science, and history meshed with each other. I did not have to analyze the act of writing of history as a holistic activity. I could think of it as such more naturally. In the production of another book on animals over 2003 and 2004 – this time about late nineteenth- and early twentieth-century working horses and called *Horses in Society: A Story of Animal Breeding and Marketing Culture, 1800–1920* – I saw, with ever greater astonishment, how much both science and art were composed of endless interconnected facets that often became linked with each other only through the medium of history.

I rose above the idea that quantitative research related to science or that qualitative research related to art. I used both methods to show how science learned from culture, and visa versa, in animal breeding, and that, because of this overlay, each could mask itself as the other. Purebred breeding is, for example, often considered to be based on genetics. It reflects, in fact, the primitive understanding of heredity laid down in seventeenth- and eighteenth-century horse-breeding theory as much as the effects of genetic knowledge (here normally population genetics). At the same time, purebred breeding clearly influenced the way both genetics and evolutionary biology developed. When the discipline of history is approached in this holistic fashion, the pageant of human thinking seemed to me to present a tapestry made of infinitely rich texture.

"Contemplation is an attempt at transmigration," wrote Ortega. He meant that thinking is the attempt to understand things, and this act he saw as love. This form of love, however, results from many interacting forces – all of which arise out of the living of a life. Individual perceptions concerning time, nostalgia, memory, art, science, beauty, truth, and the relations of each to the other dictate how we will love and therefore how we will learn. We do not see how these factors interact easily, partially because strands of light come to us over time, at different times, and in no particular order. Therefore, it can take years to be able even to articulate the complexity underlying that thought. That is the way a pathway works. I was not ready, for example, to understand well what others before me had said about these issues (and most particularly in the context of their relationship to each other) before I reached a certain point on my pathway. When I was able to do so, however, it was a strange and delicious feeling to learn that many had pondered over the meaning of these problems and had come to the same conclusions I had. Ideas about nostalgia, themes in the dinosaur paintings, and issues behind my concern for history and memory had all commanded the attention of people over the centuries. Each one expressed the thoughts differently, but the underlying concerns were usually the same.

Because the meaning of time, for example, had been a central thread running through discoveries made on my pathway into Mnemosyne's world, St Augustine's words on the subject resonated with me. He had this to say, in his *Confessions*: "What, then, is time? I know well enough what it is, provided that no one asks me; but if I am asked what it is and try to explain, I am baffled." When I read these words, I thought, "I know what you mean!" Having pondered over the subject endlessly, at one point I had written, in total confusion: "And how can one, or should one, define time?" St Augustine said something that I could interact with because of years spent thinking about the meaning of time. I wonder if his words would be so enlightening to me if had not yet gone down a pathway that led to the contemplation of time's meaning, and if I had been unfamiliar with the work of Mnemosyne and her daughters, the muses?

It is the interaction of each individual's thought while on a pathway with Mnemosyne and the muses that is critical, and it is that connection which forms the basis of my story. Depiction of a pathway can illustrate how "lack of concealment" evolves – and how it evolved in me. That is the essential beauty of life: pathways last for all consciousness, and so we need never stop learning to learn better, and never stop relating our lives to the life outside us in an enriched way. The evolving knowledge that results from the process is, in Ortega's words, "accompanied by a certain pleasurable reaction of the senses." This reaction can be called an appreciation of truth, beauty, love, or lack of concealment. They are all one, and understanding their meaning defines our humanity. Because my pathway had, in essence, been my teacher, it had, in the end, shaped my life. My pathway, then, does much to explain who and what I am. No, I suddenly thought, I would not have appreciated history more if I had gone into the doctorial program in 1970. And no, I would not have understood art better if I had gone to art school. I might well have been a better painter technically and, probably, I would have been. But I am not so sure about the heart behind the act of painting, the meaning in painting. I had to learn some of these things via a pathway.

Perhaps Plato should have the last word. In the *Symposium* he explains how people come to learn over time, or, I might say, how a pathway leads to ever-widening understanding. After someone explores avenues of knowledge and its contingent beauty, he said, eventually that person will no longer see just fragments of reality. He then expanded on how his theory concerning the interconnection of love, beauty, and knowledge worked. The individual who has reached this level will be confronted with wholeness, which is truly the "vast sea of beauty, and in gazing upon it [his/her] boundless love of knowledge becomes the medium in which [he/she] gives birth to plenty of beautiful, expansive reasoning and thinking, until [he/she] gains enough energy and bulk there to catch sight of a unique kind of knowledge whose natural object" is the greatest type of beauty. "Anyone who has been guided and trained in the ways of love up to this point, who has viewed things of beauty in the proper order and manner, will now approach the culmination of love's ways and will suddenly catch sight of something of unbelievable beauty." A person who reaches this part of a pathway will experience the wonders of truth, and will, in the process, appreciate the timelessness of things that are eternal, that do not "come to be or cease to be." That person will also be appreciating reality holistically.

Further Reading

This book touches on many subjects, and readers might find it interesting to pursue some themes in more depth. I therefore thought it would be helpful to include a note on sources. I list here not just some of the materials I referred to in the text but other useful avenues as well which lead into the issues that arise in this book. I have organized the sources to reflect the direction I followed in my pathway. Basically, I interwove my depiction of memory, contemplation, and art with a three-pronged study that looked into the philosophy of art, science, and history; the historical development of art, science, and history; and at art, science, and history from within.

The following books provided me with the most valuable sources for information on these subjects.

A fine overview on the philosophy of art and beauty can be found in Albert Hofstadter and Richard Kuhns, eds., *Philosophies of Art and Beauty: Selected Readings in Aesthetics from Plato to Heidegger* (Chicago: University of Chicago Press, 1964), and many of the quotes I provide come from this source. From this book it is possible to explore more work done by the various writers included in the volume. It is usually useful to do so, because translations affect the reading. A good example of this phenomenon can be seen when Plato's *Symposium* is read in this book and then compared with other renditions of his work. The *Symposium* is also well worth reading in its entirety. So is more of Nietzsche's work – *Ecco Homo*, for example, and *Thus Spoke Zarathustra*. See, as well, more on Goethe's thought in John Gage, ed., *Goethe on Art* (Los Angeles: University of California Press, 1980). One can range out as well to the writings of Aristotle, Plotinus, Ficino, Augustine, Shaftesbury, Kant, Schelling, Hegel, Schopenhauer, Croce, Dewey, and Heidegger. However, *Philosophies of Art and Beauty* does not include Jose Ortega y Gasset's work, which is pure delight to read, especially his first book, *Meditations on Quixote*, written in 1914. Gaston Bachelard, *The Poetics of Space* (Boston: Beacon Press, 1969), also not included, should be looked at too. J. Fisher, ed., *Essays on Aesthetics: Perspectives on the Work of M.C. Beardsley* (Philadelphia: Temple University Press, 1983), is another useful collection, in particular the article "Kant on Experiencing Beauty." The journal *British Journal of Aesthetics*, with issues published from 1960 to the present, should be consulted. Present-day attitudes, philosophically speaking, to aesthetics have become so entangled with ideas about the meaning of language, symbolism, semiotics, psychobiology, sociology, and psychology that crossover into entirely different subjects often seems to take place. I do not, therefore, list more of the modern writings on the subject. See, for example, E.H. Gombrich's excellent look at Freud and art in "Freud's Aesthetics" in *Reflections on the History of Art*, edited by R. Woodfield (Oxford: Phaidon, 1987).

The philosophy of science is a rich and enormous field, which attracts scholars interested not just in that subject but also in how it relates to a bewildering array of other human endeavours, such as psychology and medical ethics. Particularly important to the philosophy of science is Kant's *Critique of Pure Reason* (extracts cannot be found in *Philosophies of Art and Beauty*). This work is essential reading to see how all Western thought developed over the nineteenth century. Other useful works, from the point of view in particular of my book, are J. Faye, *Rethinking Science: A Philosophical Introduction to the Unity of Science* (Copenhagen: Ashgate, 2002) and A.L. Fisher and G.B. Murray, eds., *Philosophy and Science as Modes of Thinking: Selected Essays* (New York: Appleton-Century Crofts, 1969); H.P. Rickman, *The Challenge of Philosophy* (London: World Scientific, 2000); L. Schafer, *In Search of Divine Reality: Science as a Source of Inspiration* (Fayetteville: University of Arkansas Press, 1997); and T.S. Kuhn, *The Structure of Scientific Revolutions*, 2nd edition (Chicago: University of Chicago Press, 1970). Particularly interesting is I.T. Frolov, *Philosophy and History of Genetics: The Inquiry and the Debates* (London: Macdonald, 1991). The philosophy of mathematics, another huge discipline, is interconnected to that of science. Especially interesting to read, with respect to the meaning of mathematics and its relationship to human thought, are P. Benacerraf and H. Putnam, eds., *Philosophy of Mathematics* (Englewood Cliffs, N.J.: Prentice-Hall, 1964); C.J. Keyser,

Mathematics as a Culture Clue and Other Essays (New York: Scripta Mathematica, 1947); and R.E. Micens, ed., *Mathematics and Science* (London: World Scientific, 1990). See, as well, C.J. Keyser, *The Pastures of Wonder: The Realm of Mathematics and the Realm of Science* (New York: Columbia University Press, 1929). Brian Baigrie has edited two significant works: one is a four-volume set, *History of Modern Science and Mathematics* (New York: Charles Scribner's Sons, 2002); the other, *Picturing Knowledge: Historical and Philosophical Problems Concerning Science* (Toronto: University of Toronto Press, 1996), shows how the philosophy of science can overlap with both history and art. See, in particular, Michael Ruse, "Are Pictures Really Necessary? The Case of Sewell Wright's Adaptive Landscapes." This source also indicates how wide ranging the philosophy of science can be and serves as a critical introduction to the interface between art and science, a subject on which more will said later.

The philosophies of science and mathematics are directly interconnected with, and therefore lead into, philosophies of history. For the presence of historical thinking in the ideas of a philosopher/physicist, see Ernst Mach, *Die Mechanik*, 1st edition 1883, English edition known as *The Science of Mechanics: A Critical and Historical Account of its Development* and *History and the Root of the Principle of the Conservation of Energy* (1911; reprinted LaSalle, Ill.: The Open Court Publishing Company, 1960). Mach's work is of supreme importance to the whole story of history, art, and science. For more on him and for information on other philosophers as well, see J.J. Kockelmans, ed., *Philosophy of Science: The Historical Background* (New York: The Free Press, 1968). Many other books are significant and relate more purely to the philosophy of history. Ortega had many interesting things to say about the philosophy of history, especially in *History as a System* and *What Is Philosophy?* both published by W.W. Norton in New York after his death in 1955. See also Sidney Hook, ed., *Philosophy and History* (New York: New York University Press, 1963). Important to all thought covered in my book are the works of Wilhelm Dilthey, a historian and philosopher. See H.A. Hodges's two studies of Dilthey: *Wilhelm Dilthey, An Introduction* (London: Routledge, 1944) and *The Philosophy of Wilhelm Dilthey* (London: Routledge, 1952). See, as well, H.P. Rickman, *Meaning in History: Dilthey's Thoughts on History and Society* (London: Allen and Unwin, 1961) and T. Plantinga, *Historical Thinking in the Thought of Dilthey* (Toronto: University of Toronto Press, 1980). To round out a sense of the metaphysics of history, readers should also look at Georg Hegel, *The Philosophy of History* (reprinted London: Routledge and K. Paul, 1974), and R.G. Collingwood, *The Idea of History* (reprinted Oxford: Clarendon Press, 1962). The philosophies of art, science, mathematics, and history are all enormous fields, and much more is available for anyone who wishes to pursue them in greater depth. Many scholarly journals serve each of these areas and deal with specific problems.

I approached the subjects of art, science, and history from two other angles: from the point of view of development over time (or history) and from within the world of each. The historical development of art, or the history of art, is a huge and highly specialized field. For art history generally, see Francis Haskell, *History and Its Images: Art and the Interpretation of the Past* (New Haven: Yale University Press, 1993); M. Cheetham, *Kant, Art, and Art History: Moments of Discipline* (New York: Cambridge University Press, 2001); and M. Cheetham, ed., *The Subjects of Art History: Historical Objects in Contemporary Perspective* (New York: Cambridge University Press, 1998). Sir Ernst Gombrich was an important art historian and a prolific writer. His key books are *The Story of Art* (1950; reprinted London: Phaidon Press Limited, 1995) and *Art and Illusion* (1960; reprinted Princeton: Princeton University Press, 2000). Another particularly good work is Harold Spencer, ed., *Readings in Art History*, 2nd edition, 2 volumes (New York: Charles Scribner's Sons, 1976). Herschel Chipp, *Theories of Modern Art* (Berkeley: University of California Press, 1968), is also excellent. See, as well, R. Woodfield, ed., *Reflections on the History of Art* (Oxford: Phaidon, 1987). It is worth looking at Victor Burgin, *The End of Art Theory* (London: MacMillan, 1986) to understand the evolution of modern approaches within the art history discipline.

Art history attracts historians who write about the history of ideas, and here Isaiah Berlin's work is especially useful because it provides a sense of overview. See, in particular, *The Roots of Romanticism* (Princeton: Princeton University Press, 1999) and *Against the Current: Essays in the History of Ideas* (Princeton: Princeton University Press, 1959). Berlin not only shows how the motivation behind art production relates to that behind scientific research but also explains how these attitudes can change over time.

While the subject of what forms the actual demarcation between art and science has not commanded much attention from scholars, some good work focuses on the problem. I made special use of Martin Kemp, *Visualizations* (Los Angeles: University of California Press, 2000). Another book by Kemp that relate to the problem of art and science is *The Science of Art: Optical Themes in Western Art from Brunelleschi to Seurat* (New Haven: Yale University Press, 1990, 1992). See also David Topper's "Epistemology of Scientific Illustration" in *Picturing Knowledge* (listed earlier), and also two of his articles: one in E. Garber, ed., *Beyond History of Science* (Cranbury, N.J.: Associated University Presses, 1990) - a book well worth looking at in its entirety - and the other "The Parallel Fallacy: On Comparing Art and Science", *British Journal of Aesthetics* 30 (1990): 311-318. This journal has many other good articles to look over with respect to the meaning of art in a historical context. See, as well, with respect to the direct problem of art and science, D. Dutton and M. Krausz, eds., *The Concept of Creativity in Science and Art* (The Hague: Martinu Nijhoff Publishers, 1981). Stephen Jay Gould, *The Hedgehog, the Fox, and The Magister's Pox: Mending the Gap Between Science and the Humanities* (New York: Harmony Books, 2003), looks at unity issues between the humanities and science. For dinosaurs and art, see S.M. Czerkas and D. Glut, *Dinosaurs, Mammoths, and Cavemen: The Art of Charles R. Knight* (New York: E.P. Dutton, 1982), and *Dinosaurs: Past and Present*, 2 volumes (Los Angeles: Natural History Museum, 1987). For art and science in shells, see Bert Van de Roemer, "Neat Nature: The Relation Between Nature and Art in a Dutch Cabinet of Curiosities from the Early Eighteenth Century," and E.C. Spary, "Scientific Symmetries," both in *History of Science* 42 (2004): 47-84, 1-46.

The history of science commands a great deal of interest from scholars and overlaps with the philosophy of both science and mathematics. I will suggest only a few sources from this very large field. See R.M. MacLeod, *The "Creed of Science" in Victorian Britain* (Burlington, Vermont: Ashgate Variourum, 2000); H. Kragh, *An Introduction to the Historiography of Science* (Cambridge: Cambridge University Press, 1987); E. Knobloch and D.E. Row, eds., *The History of Modern Mathematics*, Volume 3: *Images, Ideas and Communities* (London: Academic Press, 1994). Specifics in the history of science are often more interesting to read than books devoted to the general history of science.

With that in mind, see Daniel Kevles, *The Physicists: The History of a Scientific Community in America* (1971; Cambridge, Mass.: Harvard University Press, 1995); Arthur Berry, *From Classical to Modern Chemistry: Some Sketches of its Historical Development* (Cambridge: Cambridge University Press, 1954); Peter Bowler, *The Mendelian Revolution: The Emergence of Hereditarian Concepts in Modern Science and Society* (Baltimore: Johns Hopkins University Press, 1989); and Bowler, *The Eclipse of Darwinism: Anti-Darwinian Evolution Theories in the Decades around 1900* (Baltimore: Johns Hopkins University Press, 1983); John Wilford, *The Riddle of the Dinosaur* (New York: Alfred A. Knopf, 1986); and James L. Powell, *Night Comes to the Cretaceous* (New York: W.H. Freeman, 1998). For a particularly beautiful book, see S. Peter Dance and David Heppell, *Classic Natural History Prints: Shells*

(London: Studio Editions, 1991). See, as well, S. Peter Dance, *A History of Shell Collecting* (Leiden: E.J. Brill–Dr. W. Blackburg, 1986). Pauline Mazumdar, *Species and Specificity* (Cambridge: Cambridge University Press, 1995), is most useful for my book.

Stephen Jay Gould, *The Structure of Evolutionary Theory* (Cambridge, Mass.: The Belknap Press of the Harvard University Press, 2002), is a magnificent and lengthy work on the evolution of the historical aspects of scientific thinking. Gould's output was enormous, and many other books written by him, not listed here, deal with the interconnected subjects of science, history, and art. He often approaches his topics from a personal point of view by looking at the subject in relation to his own experience. An interesting assessment of the historical development of astronomy in relation to that of geology can be found in Martin Gorst, *Measuring Eternity: The Search for the Beginning of Time* (New York: Broadway Books, 2001).

I looked at the history of history and at history from the historian's point of view. The historian's study of the discipline is called historiography, and while this study is closely allied with the philosophy of history and with the history of history, it can also be viewed as a separate entity. This area of study is also huge and specialized, and, consequently, there are many interesting books on both the history of history and historiography. Works focus on general patterns, on geographic (by country or region) and periodization (eras of time) approaches to history. A particularly enlightening book on the general history of history is Anthony Grafton, *The Footnote* (Cambridge: Harvard University Press, 1997). For more on the history of history and on historiography from an overview point of view, see Ernst Breisach, *Historiography: Ancient, Medieval and Modern* (1983; Chicago: University of Chicago Press, 1994). (This book also contains a magnificent bibliography on the history of history and historiography.) Benedetto Croce, *Theory and History of Historiography* (London: George Harrap, 1921), is a classic. Other works worth pursuing are H.

Butterfield, *The Origins of History* (New York: Basic Books, 1981); E.H. Carr, *What Is History?* (1967; reprinted Bassingstoke, Britain: Palgrave, 2001); G. Himmelfarb, *The New History and the Old: Critical Essays and Reappraisals* (reprinted Cambridge, Mass.: Harvard University Press, 2004); E.H. Gombrich, *In Search of Cultural History* (Oxford: Clarendon Press, 1969); and R. Jann, *The Art and Science of Victorian History* (Columbus, Ohio: Ohio State University Press, 1985). Jacques Le Goff, *History and Memory* (New York: Columbia University Press, 1991), and Raphael Samuel, *Theatres of Memory* (New York: Verso, 1994), extend the subjects of the history of history and historiography into a more general look at history's interface with aspects of popular culture. A few other sources I found especially useful are G.A. Reisch, "Chaos, History, and Narrative," *History and Theory* 30 (1991): 1–20, and, in the same volume, D.N. McCloskey, "History, Differential Equations, and the Problem of Narration," 21–36. See also Ian Winchester, "History, Scientific History and Physics," *Historical Methods* 17 (1984): 95–106.

The realm of science is, as everyone knows, truly staggering in its size. I looked at this world from within by pursuing specific fields that were not part of immunology. I studied areas of science that related to paleontology. Anyone interested in anatomy, biology, or paleontology will find the sources I list below to be fascinating. Sisson and Grossman's *The Anatomy of Domestic Animals*, volumes 1 and 2, edited by R. Getty, 5th edition (London: W.B. Saunders, 1975); George C. Kent, *Comparative Anatomy of the Vertebrates*, 6th edition (St Louis: Times Mirror/Mosby College Publishing, 1987); A.S. Romer and T.S. Parsons, *The Vertebrate Body*, 6th edition (New York: Saunders College Publishing, 1986); and Henry Gray, *Gray's Anatomy* (c. 1850; reprinted New York: Bounty Books, 1977) are all excellent. I also read scientific articles on dinasaurian life. A few, just as examples, include Barnum Brown, "A New Trachodont Dinosaur, Hypacrosaurus, from the Edmonton Cretaceous of Alberta," *American Museum of Natural*

History 32 (1913): 395–406; David B. Weishampel, "The Nasal Cavity of Lambeosaurine Hadrosaurids (Reptilia: Ornithischia): Comparative Anatomy and Homologies," *Journal of Paleontology* 55 (1981): 1046–57; and James A. Hopson, "The Evolution of Cranial Display Structures in Hadrasaurian Dinosaurs," *Paleobiology* 1 (1975): 21–43. An interesting book from another area of biology is Geerat J. Vermeij, *A Natural History of Shells* (Princeton: Princeton University Press, 1993).

I approached art from within, in a triple fashion. First, I learned how to produce art by studying the techniques of painting and drawing. (I will not list here the sources I used for the actual production of art, except to mention the magnificent book by Johannes Itten, *The Art of Color* (New York: Nostrand Reinhold, 1961), and, for historical reasons, D.A. Anfam et al, *Techniques of the Great Masters of Art* (Secaucus, NJ: Chatwell Books, 1985). Second, I read what artists had to say about the visual arts. Art can be appreciated most fully – by a non-painter – when seen through the eyes of painters. Leonardo Da Vinci, *Notebooks* (reprinted New York: Black Dog & Leventhal, 2005), make fascinating reading. Lectures by Joshua Reynolds, found in *Discourses on Art* (reprinted New Haven: Yale University Press, 1975), are magnificent. Robert Henri, *The Art Spirit* (1923; reprinted New York: Harper & Row, 1958, Icon Edition, 1984) is inspiring. John Ruskin, *The Elements of Drawing* (1857; reprinted New York: Dover Publications, Inc;, 1971) is essential for anyone interested in the process of art. Vincent Van Gogh's letters are beautiful (these have been published under many formats). A good collection of artists' thoughts can be found in Robert Goldwater and Marco Treves, eds., *Artists on Art* (New York: Pantheon Books, 1945). The poetry of Michelangelo and of the Pre-Raphaelite painters adds another dimension to the vision of painters.

Third, I produced art, and in the end executed many hundreds of paintings. Today some of these works are owned by private collectors and can also be found in the following public collections:

Blake, Cassels & Graydon; Domco Foodservices; Goodman and Goodman; McCarthy Tetrault; Queenston Goldmines Ltd.; PriceWaterhouseCoopers; Bell Canada; Bishop Strachan School; and Upper Canada College. I held solo exhibitions at Kinsman Robinson Galleries in Toronto, September 1991 and September 1987; Gallery Gabor Limited in Toronto, September 1985; and Gustafsson Galleries in Toronto, October 1983 and September 1982. I took part in group exhibitions at Kinsman Robinson Galleries, December 1989 and September 1988; City Hall of Etobicoke, January 1988; and Gallery Gabor Limited, summer 1986. Several articles appeared on my art: Ron Starr, "Margaret Derry," *Ontario Living*, March 1986; and Sarah Yates, "Corporations as Collectors, Getting into the Art Picture," *Toronto Office Space Guide*, winter 1985.

I looked at history from within, not just by a study of historiography but also through the pursuit of history itself. I gave talks on my historical research, for colloquia at universities in both Canada and the United States, the most recent being at Yale University in January 2005. I also spoke in China at the International Congress for the History of Science in July 2005. My general areas of history are Western agriculture, livestock production, the international problem of animal disease, and the culture of animal breeding over the last two hundred years. Of particular concern to me is the fusion of science and culture in the evolution of purebred breeding.

I wrote history, both in the form of books and articles. I mentioned the books in this story, but I will list them here in full: *Ontario's Cattle Kingdom: Purebred Breeders and Their World, 1870–1920* (Toronto: University of Toronto Press, 2001); *Bred for Perfection: Shorthorn Cattle, Collies, and Arabian Horses Since 1800* (Baltimore: Johns Hopkins University Press, 2003); and *Horses in Society: A Story of Animal Breeding and Marketing Culture, 1800–1920* (Toronto: University of Toronto Press, 2006). Selected articles include "Patterns of Gendered Labour and the

Development of Ontario Agriculture," in E. Montigny and L. Chambers, eds., *Ontario Since Confederation: A Reader* (Toronto: University of Toronto Press, 2000); "Gender Conflicts in Dairying: Ontario's Butter Industry, 1880–1920," *Ontario History*, volume LXXXX, 1998: 31-47 and "Contemporary Attempts to Understand the Cattle Plague of 1865," Victorian Studies Association, *Ontario Newsletter*, fall 1994.

Looking at the philosophy of art, science, and history, at the historical development of art, science, and history, and then at art, science, and history from within (more particularly, in art through painting, in science through reading paleontology, and in history through the writing of history) all helped me to relate better to the wholeness found in literature. I came to see deeper meanings in the many masterpieces of literature and poetry I had read over the years. I referred in this book to a number of novels that I particularly enjoyed. These examples serve only as a few of the inexhaustible riches that can be found in the world of literary writing. Journals can be extremely revealing. A bibliographic list of meaningful literature, then, would be too extensive to include here.

It will be evident from this short bibliography that each area I refer to in this book is, in itself, an enormous field. Does it make sense merely to touch on the subjects in the way I do? Is my simplification so stripped of information emanating from these fields that the result is naïvety? That is for the reader to decide. It seems to me, though, that, with no attending directional force, the idea of wholeness itself can be lost. When a mere overview or synthesis of the magnificent knowledge existing in all these endeavours of humanity alone is offered, it is not enough to explain wholeness. That is, perhaps, not a new thought. Efforts to perceive how wholeness works have been tried without the pure synthesis that would reduce the disciplines to overarching statements.

The subject of art's linkage to science, for example, has made scholars from the arts and from the sciences explore the unity by methods other than the writing of books. Art exhibits have been put together. For example, "Spectacular Bodies: The Art and Science of the Human Body from Leonardo to Now," an exhibition originating at the Hayward Gallery in London, England, over 2000 and 2001, explored the way the human body, from a medical point of view, has been used to make art. Anatomical investigations were presented as art. The Wellcome Trust, which funds studies relating to medicine, has set up "SciArt," a scheme designed to support work done by artists and scientists that would bring them closer together. Physicists are also interested in the union. F. David Peat, who stated that "science and art are two different approaches that complement each other, and [both] are needed to produce a balanced vision of the world," runs a forum on the Internet to discuss art and science. Many of these art/science endeavours, however, do not delve into the deeper heart of the matter, a fact recognized by some. Shared motives, illustrations using science, do not necessarily demonstrate penetrating aesthetic vision. The move of some scientists to support the value of art can also arouse the ire of other scientists, who see knowledge for the good of humanity in science as outstripping that in art. And many believe that art makes no effort to bring science into its thinking.

Perhaps the voyage of a pathway, which is what this book is about, will help, at least in a small way, to heal the cleavage of which so many scholars are aware.

Index